Mike McGrath

HTML

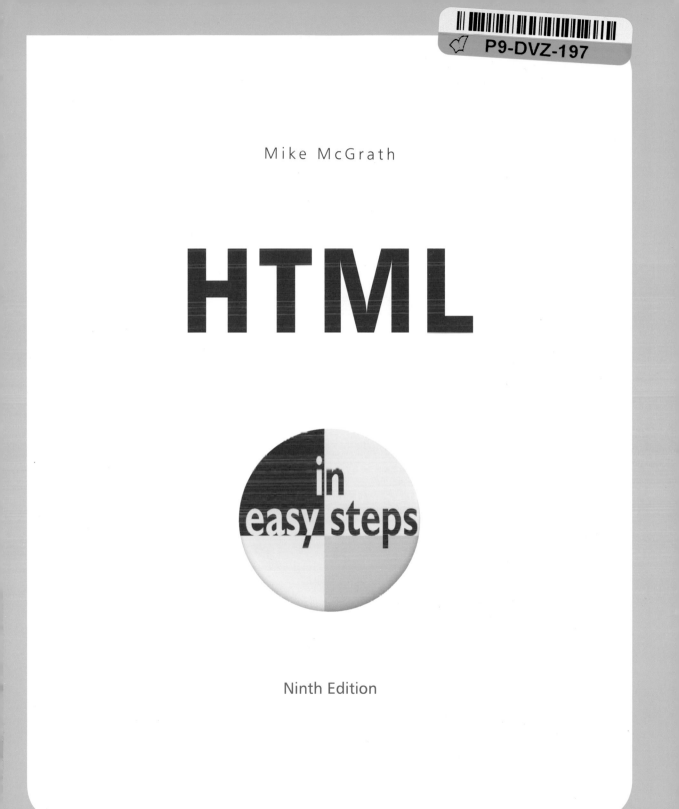

Ninth Edition

In easy steps is an imprint of In Easy Steps Limited
16 Hamilton Terrace · Holly Walk · Leamington Spa
Warwickshire · United Kingdom · CV32 4LY
www.ineasysteps.com

Ninth Edition

Notice of Liability
Every effort has been made to ensure that this book contains accurate
and current information. However, In Easy Steps Limited and the
author shall not be liable for any loss or damage suffered by readers
as a result of any information contained herein.

Trademarks
All trademarks are acknowledged as belonging to their respective
companies.

In Easy Steps Limited supports The Forest Stewardship Council (FSC),
the leading international forest certification organization. All our titles
that are printed on Greenpeace approved FSC certified paper carry the
FSC logo.

MIX
Paper from
responsible sources
FSC® C020837

Printed and bound in the United Kingdom

ISBN 978-1-84078-876-1

Contents

How to Use This Book

The examples in this book demonstrate HTML features that are supported by modern web browsers, and the screenshots illustrate the actual results produced by the listed code examples. Certain colorization conventions are used to clarify the code listed in the steps...

HTML tags are colored blue, literal text is colored black, and comments are colored green. Element attribute values are colored orange in HTML, in CSS, and in JavaScript code:

```
<!-- The Traditional Greeting. -->
<h1 id="info">Hello World!</h1>
```

CSS style sheet code is colored blue, and values assigned to properties are colored red:

```
info { color : Red ; background : Yellow ; }
```

JavaScript code is colored blue, programmer-specified names are colored red, literal text is colored black, and code comments are colored green:

```
let msg = 'My Web Page' ; document.getElementById( 'info' ).innerText = msg
```

Additionally, in order to identify each source code file described in the steps, a file icon and file name appears in the margin alongside the steps:

page.html   styles.css   functions.js   chart.pdf   vector.svg   audio.mp3   echo.py

You can download a single ZIP archive file containing all the complete example files that appear in this book by following these easy steps:

1   Browse to **www.ineasysteps.com** then navigate to Free Resources and choose the Downloads section

2   Next, find HTML in easy steps, 9th edition in the list, then click on the hyperlink entitled All Code Examples to download the ZIP archive file

3   Now, extract the archive contents to any convenient location on your computer

If you don't achieve the result illustrated in any example, simply compare your code to that in the original example files you have downloaded to discover where you went wrong.

1 Get Started in HTML

This chapter is an introduction to the exciting world of HTML. It demonstrates how to create a valid HTML document and how to include style rules, script code, and linked resources.

Meet HTML

Historically, the desire to have text printed in specific formats meant that original manuscripts were "marked up" with annotation to indicate to the book printer how the author would like sections of text laid out. This annotation had to be concise and needed to be easily understood both by the printer and the author. A series of commonly-recognized abbreviations therefore formed the basis of a standard markup language.

HyperText Markup Language (HTML) is a modern standard markup language that uses common abbreviations called "tags" to indicate to the web browser how the author would like to have sections of a web page laid out. It was first devised in 1989 by British physicist Tim Berners-Lee at CERN in Switzerland (the European organization for nuclear research) to share all computer-stored information between the CERN physicists. Berners-Lee created a text browser to transfer information over the internet using hypertext to provide point-and-click navigation. In May 1990 this system was named the World Wide Web and was enhanced in 1993 when college student Marc Andreessen added an image tag. Now that HTML could display both text and images, the World Wide Web quickly became hugely popular.

As various web browsers were developed, their makers began to add individual proprietary tags – effectively creating their own versions of HTML! The World Wide Web Consortium (W3C) standards organization recognized the danger that HTML could become fragmented, so they created a standard specification to which all web browsers should adhere. This successfully encouraged the browser makers to support the standard tags. The final W3C standard specification of HTML5 is now continued by the Web Hypertext Application Technology Working Group (WHATWG) as the "HTML Living Standard".

The World Wide Web comprises a series of large-capacity computers, known as "web servers", which are connected to the internet via telephone lines and satellites. The web servers each use the HyperText Transfer Protocol (HTTP) as a common communication standard to allow any computer connected to any web server to access files across the web.

HTML web pages are merely plain text files that have been saved with a ".htm" or ".html" file extension, such as **index.html**.

You can find the HTML Living Standard specification, and other related specifications, online at **whatwg.org**

In order to access an HTML file across the internet, its web address must be entered into the address field of the web browser. The web address is formally known as its "Uniform Resource Locator" (URL), and typically has three parts:

- **Protocol** – any URL using the HTTP protocol begins by specifying the protocol as **http://** or secure **https://**

- **Domain** – the host name of the computer from which the file can be downloaded. For instance: **www.example.com**

- **Path** – the file name prefixed by any parent directory names where applicable. For instance: **/htdocs/index.html**

A URL describing the location of a file by protocol, domain, and path is stating its full "absolute address". Files resident within the same domain can be referenced more simply by their "relative address", which means that files located in the same directory can be referenced just by their file name. Additionally, a relative address can reference a file in its parent directory by prefixing its name with "../". For instance, a file named "higher.html" in the parent directory can be referenced as **../higher.html**.

How do web servers work?
When you enter a URL into the browser address field, the browser first examines the protocol. Where the protocol is specified as HTTP, or assumed to be HTTP if unspecified, the browser recognizes that a file is being sought from a web server. It then contacts a Domain Name Server (DNS) to look up the numerical Internet Protocol (IP) address of the specified domain name. Next, a connection is established with the web server at that IP address to request the file at the specified path. When the file is successfully located, it is copied back to the browser, otherwise the web server sends an error code, such as "404 – Page Not Found".

A successful response sends HTTP headers to the web browser, describing the nature of the response, along with a copy of the requested file. The HTTP headers are not normally visible but can be examined using various development tools, such as the F12 Developer Tools feature in the Google Chrome web browser.

Understand Structure

The skeletal structure of an HTML document has three parts:

- **Document type declaration** – declaring precisely which version of HTML is used to mark up the document.

- **Head section** – providing descriptive data about the document itself, such as the document's title and the character set used.

- **Body section** – containing the content that is to appear when the document gets loaded into a web browser.

Document type declaration

The document type declaration must appear at the start of the first line of every HTML document to ensure the web browser will "render" (display) the document in "Standards Mode" – following the HTML specifications. The document type declaration tag for all HTML documents looks like this:

<!DOCTYPE HTML>

It is important to note that HTML is not a case-sensitive language – so the document type declaration tag, and all other tags, may alternatively be written in any combination of uppercase and lowercase characters. For example, the following are all valid:

<!DOCTYPE html>

<!Doctype Html>

<!doctype html>

The choice of capitalization is yours, but it is recommended you adhere consistently to whichever style you choose. The document type declaration tag capitalization style favored throughout this book uses all uppercase to emphasize its prominence as the very first tag on each page – but all other tags are in all lowercase.

Those familiar with earlier versions of HTML may be surprised at the simplicity of the HTML document type declaration. In fact, the document type declaration in earlier versions was not actually part of the HTML language – so required lengthy references to schema documents. By contrast, the modern HTML document type declaration is an intrinsic part of HTML itself.

Hot tip

The document type declaration in earlier versions of HTML was part of the Standard Generalized Markup Language (SGML) from which HTML is derived.

...cont'd

The entire document head section and body section can be enclosed within a pair of **<html> </html>** tags to contain the rest of the document. The HTML specification actually states that these are optional, but it is logical to provide a single "root" element. Most HTML tags are used in pairs like this to act as "containers" with the syntax **< tagname > data </ tagname >**.

Head section
The document's head section begins with an HTML opening **<head>** tag and ends with a corresponding closing **</head>** tag. Data describing the document can be added later between these two tags to complete the HTML document's head section.

Body section
The document's body section begins with an HTML opening **<body>** tag and ends with a corresponding closing **</body>** tag. Data content to appear in the browser can be added later between these two tags to complete the HTML document's body section.

Code comments
Comments can be added at any point within both the head and body sections between a pair of **<!--** and **-->** tags. Anything that appears between the comment tags is ignored by the browser.

Fundamental structure
So, the markup tags that create the fundamental structure of every HTML document look like this:

```
<!DOCTYPE HTML>

<html>
 <head>
  <!-- Data describing the document to be added here. -->
 </head>

 <body>
  <!-- Content to appear in the browser to be added here. -->
 </body>

</html>
```

An HTML "element" is any matching pair of opening and closing tags, or any single tag not requiring a closing tag – as described in the HTML element tags list on the inside front cover of this book.

The "invisible" characters that represent tabs, newlines, carriage returns, and spaces are collectively known as "whitespace". They may optionally be used to inset the tags for clarity.

Create Documents

The fundamental HTML document structure described on page 11, can be used to create a simple HTML document in any plain text editor – such as Windows' Notepad application. In order to create a valid "barebones" HTML document, information must first be added defining the document's primary written language, its character encoding format, and its title.

The document's primary language is defined by assigning a standard language code to a **lang** "attribute" within the opening **<html>** root tag. For the English language the code is **en**, so the complete opening root element looks like this: **<html lang="en">**.

The document's character encoding format is defined by assigning a standard character-set code to a **charset** attribute within a **<meta>** tag placed in the document's head section. The recommended encoding is the popular 8-bit Unicode Transformation Format for which the code is **UTF-8**, so the complete element looks like this: **<meta charset="UTF-8">**.

Finally, the document's title is defined by text between a pair of **<title> </title>** tags placed in the document's head section.

Follow these steps to create a valid barebones HTML document:

Beware

HTML documents should not be created in word processors such as MS Word, as those apps include additional information in their file formats.

HTML

hello.html

Hot tip

The **<meta>** tag is a single tag – it does not have a matching closing tag. See the element tags list on the inside front cover of this book to find other single tags.

1 Launch your favorite plain text editor then start a new document with the HTML document type declaration
<!DOCTYPE HTML>

2 Below the document type declaration, add a root element that defines the document's primary language as English
<html lang="en">
  <!-- Head and Body sections to be added here. -->
</html>

3 Within the root element, insert a document head section
<head>
  <!-- Descriptive information to be added here. -->
</head>

4 Within the head section, insert an element defining the document's encoding character set
<meta charset="UTF-8">

5 Next, within the head section, insert an element defining the document's title
<title>Get Started in HTML</title>

6 After the head section, insert a document body section
<body>
  <!-- Actual document content to be added here. -->
</body>

7 Within the body section, insert a size-one large heading
<h1>Hello World!</h1>

```
hello.html - Notepad                          —  □  ×
File  Edit  Format  View  Help
<!DOCTYPE HTML>

<html lang="en">

<head>
<meta charset="UTF-8">
<title>Get Started in HTML</title>
</head>

<body>
<h1>Hello World!</h1>
</body>

</html>
```

Hot tip

You will discover more about headings on pages 30-31.

Don't forget

The quotation marks around an attribute value are usually optional but are required for multiple values. For consistency, attribute values in the examples throughout this book are all surrounded by quotation marks.

13

8 Set the file's encoding to the UTF-8 format, then save the document as "hello.html"

```
Encoding: UTF-8                    ▼    Save
          ANSI
          UTF-16 LE
          UTF-16 BE
          UTF-8
          UTF-8 with BOM
```

9 Now, open the HTML document in a modern web browser to see the title displayed on the title bar or tab, and the document content displayed as a large heading

```
◉ Get Started in HTML        ×   +            —  □  ×
← → C ⌂  G hello.html                          😊  ⋮

Hello World!
```

Validate Documents

Just as text documents may contain spelling and grammar errors, HTML documents may contain various errors that prevent them from conforming to the specification rules. In order to verify that an HTML document does indeed conform to the rules of its specified document type declaration, it can be tested by a validator tool. Only HTML documents that pass the validation test successfully are sure to be valid documents.

Web browsers make no attempt at validation so it is well worth verifying every HTML document with a validator tool before it is published, even when the content looks fine in your web browser. When the browser encounters HTML errors it will make a guess at what is intended – but different browsers can make different interpretations so may display the document incorrectly. Conversely, valid HTML documents should always appear correctly in any standards-compliant browser.

The World Wide Web Consortium (W3C) provides a free online validator tool at **validator.w3.org** that you can use to check the syntax of your web documents:

1 With an internet connection, open your web browser and navigate to the W3C Validator Tool at **validator.w3.org** then click on the **Validate by File Upload** tab

Other tabs in the validator allow you to enter the web address of an HTML document located on a web server to "Validate by URI" or copy and paste all code from a document to "Validate by Direct Input".

W3 The W3C Markup Validation Ser ☓ +

← → C △ 🔒 validator.w3.org/#validate_by_upload ☆ 😔 ⋮

W3C® Markup Validation Service
Check the markup (HTML, XHTML, ...) of Web documents

Validate by URI   **Validate by File Upload**   Validate by Direct Input

Validate by File Upload
Upload a document for validation:

File: [Choose File] hello.html

▶ More Options

(Check)

https://validator.w3.org/#

2   Click the **Browse** button then navigate to the HTML document you wish to validate – once selected, its local path appears in the validator's "File" field

3   Next, click the validator's **Check** button to upload a copy of the HTML document and run the validation test – the results will then be displayed

Don't forget

The validator automatically detects the document's character set and HTML version.

If validation fails, the errors are listed so you may easily correct them. When validation succeeds, you may choose to include a suitable logo at the end of the document to prove validation:

Hot tip

The validation logo can be customized to describe the technology classes used by the web page. Discover the logo Badge Builder online at **w3.org/html/logo** where you can generate the code to paste into your HTML document and so display a suitable logo.

Bestow Titles

The specifications require every HTML document to have a title, but its importance is often overlooked. The document title should be carefully considered, however, as it is used extensively:

- **Bookmarks** – save the document title to link back to its URL.

- **Title Bar** – a web browser window may display the title.

- **Navigation Tab** – a web browser tab may display the title.

- **History** – saves the document title to link back to its URL.

- **Search Engines** – read the document title and typically display it in search results to link back to its URL.

Document titles should ideally be short and meaningful – each tab on a modern tabbed browser may display only 10 characters.

The specifications do not define a naming scheme for document titles but do encourage authors to consider accessibility issues in all aspects of their web page designs.

Document titles throughout a website should follow a consistent naming convention and capitalize all major words. One popular naming convention provides a personal or company name and brief page description separated by a hyphen. For example, "Amazon - C Programming in easy steps". An alternative puts the description first, so it remains visible when the title is truncated. For example, "C Programming in easy steps - Amazon".

Amazon - C Programming in eas   ✕

C Programming in easy steps - A   ✕

You can find a chart of all character entities at dev.w3.org/html5/html-author/charref

Document titles and document content may contain special characters that are known in HTML as "entities". Each entity reference begins with an ampersand and ends with a semicolon. For example, the entity **<** (less than) creates a "<" character and the entity **>** (greater than) creates a ">" character. These are often needed to avoid confusion with the angled brackets that surround each HTML tag. Other frequently used entities include ** ** (a single non-breaking space), **•** (bullet point), **©** (©), **®** (®), **™** (™), and **"** (quotation mark). These are best avoided in document titles, however, as the vocal narrator used by visually impaired viewers may read each entity character as a word.

Start a new HTML document with a type declaration
<!DOCTYPE HTML>

 Add a root element containing head and body sections
<html lang="en">

<head>
<!-- Data describing the document to be added here. -->
</head>

<body>
<!-- Content to appear in the browser to be added here. -->
</body>

</html>

title.html

Within the head section, insert a meta element specifying the character set and an empty title element
<meta charset="UTF-8">
<title> </title>

Within the title element insert a title including entities
<HTML in easy steps>

Save the document then open it in your web browser

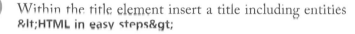

Start a vocal narrator to hear that the title may be read out as "Less-than-HTML-in-easy-steps-greater-than"

Edit the document title to make it more user-friendly
"HTML in easy steps"

"HTML in easy steps"   ✕

Save the document once more then open it in your web browser to hear the narrator now read the document title as "HTML in easy steps"

The character set can be defined in uppercase, as shown here, or in lowercase as "utf-8".

In Windows 10, press **WinKey + Ctrl + Enter** to launch the narrator, then click the tab to hear the title. Title text that is not visible on the tab will still be read by the narrator. Windows 10 ignores angled brackets in a title, but they are read literally by the narrator in earlier versions of Windows.

Supply Metadata

Meta information is simply data that describes other data. In the context of HTML, document metadata describes the document itself – rather than the document's contents.

HTML metadata is defined in the head section of the HTML document using the **\<meta\>** tag. The **\<meta\>** tag is an "empty" tag that needs no matching closing tag to create an HTML element – it is only used to specify information with its tag attributes. Previous examples have used this tag to specify the document's character-set. Further **\<meta\>** tags can be added to describe other aspects of the document.

Given the number of handheld devices that may view a web page, it is useful to optimize the page for smaller screens by including this **\<meta\>** tag in all your HTML documents' head sections:

```
<meta  name="viewport"
          content="width=device-width, initial-scale=1">
```

This will ensure your document will fill the device screen width and sets the initial zoom level so the content is not zoomed.

Setting the **width** to the **device-width** typically sets the **initial-scale** to **1** automatically, but it doesn't hurt to set it explicitly as meta data.

A **\<meta\>** tag can also assign a document HTTP header property to an **http-equiv** attribute and can specify that property's value to a **content** attribute. You can assign the HTTP "refresh" property to an **http-equiv** attribute to reload the page after a number of seconds specified to its **content** attribute – for example, to reload the page after five seconds, like this:

```
<meta http-equiv="refresh" content= "5">
```

This technique is often used on websites to dynamically update news or status items, as it does not depend on JavaScript support.

Another popular use redirects the browser to a new web page after a specified number of seconds, like this:

```
<meta http-equiv="refresh" content= "5 ; url='new-page.html' ">
```

In this case, the **\<meta\>** tag's **content** attribute specifies both the number of seconds to delay and the new URL to load.

1 Create a barebones HTML document

```html
<!DOCTYPE HTML>
<html lang="en">
<head>
<meta charset="UTF-8">
<!-- More metadata to be inserted here. -->
<title>Meta Refresh</title>
</head>
<body>
<h1>Moving in 5 Seconds...</h1>
</body>
</html>
```

refresh.html

2 Insert two more elements of metadata

```html
<meta  name="viewport"
       content="width=device-width, initial-scale=1">
<meta http-equiv="refresh"
       content= "5 ; url='https://ineasysteps.com' ">
```

3 Save the document then open it in your web browser and wait five seconds to see the browser redirect

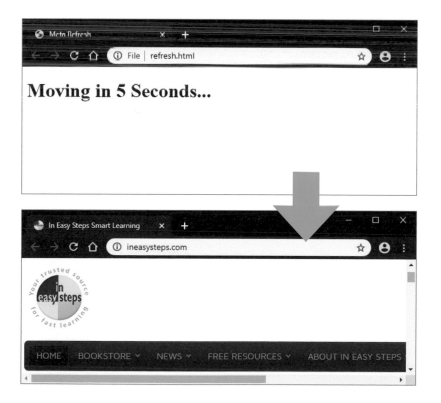

When you only specify the domain to the **url** attribute, as in this case, the browser will automatically load the **index.html** page at that domain location.

Describe Contents

In addition to specifying the document's character-set and expiry date, **<meta>** tags can be used to provide information that may be used by search engines. This offers no guarantee of high ranking, however, as search engines also use other page information for that purpose – especially the document title. Typically, a Search Engine Results Page (SERP) will show the meta description in search results below the page title.

Search Engine Optimization (SEO) is highly prized to ensure a web page will appear at the top of a SERP to increase traffic to a website. Unfortunately, there is no sure-fire technique to achieve this as the search engines constantly change the algorithm by which pages are ranked. It is, however, useful to provide metadata that describes the page content.

Descriptive **<meta>** tags have a **name** attribute that is assigned a "description" value, and a **content** attribute that is assigned a description of the page contents.

The description should be between 50-160 characters long, as lengthy descriptions may be truncated. The description should include keywords relative to the text content. For example, a search for "italian ceramics" could return all web pages with "italian" and "ceramics" in their description.

The description serves as advertising copy so a readable, compelling description using important keywords will encourage visits to the page from a SERP. You should not repeat keywords in the description, but do try to use the plural form for keywords – to match searches made with both the single and plural form of that word. Additionally, you should not include double quotation marks in the description as Google may truncate the description at a double quotation mark.

If a website contains pages of identical or very similar content, you can specify which page is to be indexed by including a "canonical link" in your HTML code to indicate the preferred source. This uses a **<link>** tag containing a **rel** (relationship) attribute to specify a "canonical" value, and an **href** (hypertext reference) attribute to specify the URL address of the preferred page.

All search engines find pages to add to their index – even if the page has never been submitted to them.

Always include the three most important keywords in the description.

1 Create a barebones HTML document

```
<!DOCTYPE HTML>
<html lang="en">
<head>
<meta charset="UTF-8">
<!-- More metadata to be inserted here. -->
<title>Tuscan Home Decor</title>
</head>
<body> <h1>Beautiful Tuscan Ceramics</h1> </body>
</html>
```

keywords.html

2 Insert a metadata description of the web page

```
<meta name="description" content="Explore our
extensive range of high quality italian ceramics including
tuscan majolica, dinnerwares, vases, plates, and bowls">
```

3 Next, in the head section, add an element to specify that this page is the preferred page for indexing purposes

```
<link rel="canonical"
        href="https://www.example.com/keywords.html" >
```

4 Save the document then visit the Chrome Web Store at **chrome.google.com/webstore/category/extensions** and search for "seo" to add a search engine analysis extension

5 Open the HTML document in the Google Chrome web browser then use the analysis tool to see the meta data

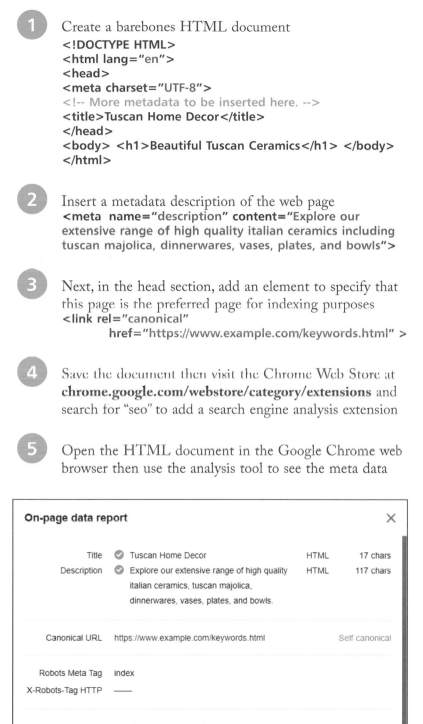

On-page data report				X
Title	✓ Tuscan Home Decor		HTML	17 chars
Description	✓ Explore our extensive range of high quality italian ceramics, tuscan majolica, dinnerwares, vases, plates, and bowls.		HTML	117 chars
Canonical URL	https://www.example.com/keywords.html			Self canonical
Robots Meta Tag	index			
X-Robots-Tag HTTP	——			
H1	Beautiful Tuscan Ceramics			

Hot tip

There are a number of free meta tag generators available online – enter "free meta tag generator" into a search engine.

Add Styles

Cascading Style Sheets (CSS) rules can be incorporated within HTML documents to control the presentational aspects of each element on the page. The use of style sheets has replaced all features of HTML that formerly related to presentation. For example, the **<font>** tag has become obsolete, as font family, weight, style, and size are now specified by a style sheet rule.

Style sheets embedded with **<style> </style>** tags can be added within the head section of an HTML document to enclose rules governing how the content will appear. For example, a simple style sheet containing rules to determine the appearance of all size-one headings could look like this:

<style>

h1 { color : red ; background : yellow ; }

</style>

This is acceptable and will validate but, in line with the aim of HTML to separate content from presentation, style sheets may be contained within a separate file. The great advantage of placing style sheets in separate files is that they can be applied to multiple HTML documents – thus making website maintenance much easier. Editing a shared style sheet instantly affects each HTML document that shares that file.

An external style sheet is incorporated within an HTML document by adding a **<link>** tag in the document's head section. This must contain a **rel** (relationship) attribute assigned a "stylesheet" value, and the URL of the style sheet must be assigned to its **href** (hypertext reference) attribute – for example, add an adjacent style sheet file named "style.css", like this:

<link rel="stylesheet" href="style.css">

You can also specify style rules "in-line" to a style attribute of presentational HTML tags, like this:

<h1 style="color:red">

In-line style rules are useful in some circumstances but can make page maintenance more difficult.

When multiple rules select the same property of an element for styling, the rule read last by the browser will generally be applied, but in-line rules take precedence over embedded rules and external rules. Embedded rules take precedence over external rules.

You can learn more about style sheet rules with the companion book in this series: **CSS in easy steps**.

1 Create a barebones HTML document

```html
<!DOCTYPE HTML>
<html lang="en">
<head>
<meta charset="UTF-8">
<title>Style Sheet Example</title>
</head>
<body>
<h1>Styled Heading</h1>
</body>
</html>
```

style.html

2 Next, in the head section, add an embedded style sheet

```html
<style>
h1 { color : Red ; background : Yellow ; }
</style>
```

3 Now, in the head section, insert a link to an adjacent external style sheet file

```html
<link rel="stylesheet" href="style.css">
```

style.css

4 Save the HTML document then open a new text editor window and precisely copy this style sheet

```css
h1
{
        border : 10px dashed Blue ;
        padding : 5px ;
        width : 500px ;
}
```

5 Save the Cascading Style Sheets file in the same directory as the HTML document, then open the web page in your browser to see the style rules applied

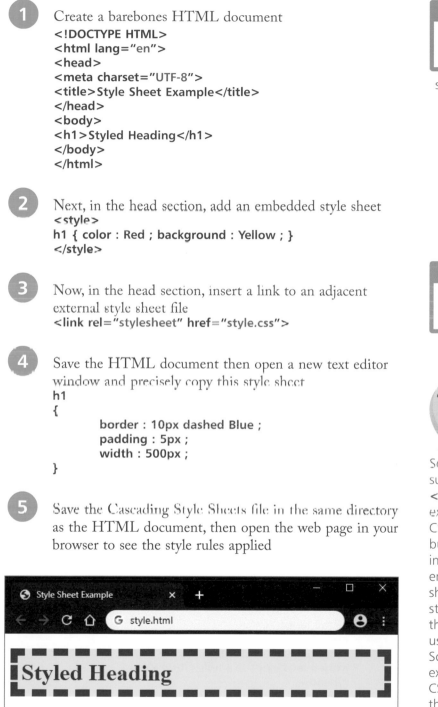

Some HTML elements, such as **<div>** and **<span>** (see page 42), exist purely for styling CSS is a separate topic but many examples in this book include embedded CSS style sheet rules to provide standalone example files that demonstrate the use of HTML elements. Some of the source code examples include unlisted CSS rules to illustrate the size and position of HTML elements and their content in screenshots.

Include Scripts

Scripts can be incorporated within HTML documents to interact with the user and to provide dynamic effects. This ability has become increasingly important with the development of pages in which sections of the page can be dynamically updated. Previously, the browser would typically request an entire new page from the web server, which was less efficient and more cumbersome.

JavaScript code enclosed by **<script> </script>** tags can be embedded within an HTML document. These are best placed in the body section of the document, just before the **</body>** closing tag, so the browser can process the content of the document before reading the script.

Remember that the **<script>** tag always needs to have a matching closing tag.

In line with the aim of HTML to separate content from presentation, scripts may also be contained in a separate file. In this case, the URL address of the script file must be assigned to a **src** attribute within the **<script>** tag. The **</script>** closing tag is also required. These, too, can be placed at the end of the body section of the HTML document, as the browser will treat the external script as if it was embedded there – for example, to add an adjacent external script file named "script.js", like this:

```
<script src="script.js"></script>
```

You can also specify script "in-line" to on-event attributes of HTML tags. For example, to recognize a mouse click event:

```
<h1 onclick="alert('Clicked!')">
```

You can learn more about scripting with the companion book in this series: **JavaScript in easy steps**.

In-line script is useful in some circumstances but can make page maintenance more difficult. Alternative fallback content can be provided in the document's body section between **<noscript> </noscript>** tags, which will only be displayed when script functionality is absent or disabled.

script.html

1 Create a barebones HTML document

```
<!DOCTYPE HTML>
<html lang="en">
<head>
<meta charset="UTF-8">
<title>JavaScript Example</title>
</head>
<body>
</body>
</html>
```

2 In the body section, insert a fallback message and heading
```
<noscript>JavaScript Is Not Enabled!</noscript>
<h1 onclick="this.innerText='Mouse Clicked!';
        this.style.color='Red'">Active Heading</h1>
```

3 At the end of the body section, add an embedded script
and nominate an external script
```
<script>
document.getElementsByTagName('h1')[0].onmouseover =
function ( ) {
this.innerText= 'Mouse Is Over' ; this.style.color='Blue' }
</script>
<script src="script.js"></script>
```

4 Save the HTML document then open a new text editor
window and create the external script
```
document.getElementsByTagName('h1')[0].onmouseout =
function ( ) {
this.innerText= 'Active Heading' ; this.style.color = 'Black' }
```

script.js

5 Save the JavaScript file as "script.js" in the same directory
as the HTML document, then open the web page in your
browser and click on the heading

Don't forget

Some HTML elements, such as **<template>** and **<slot>** (see page 116), exist purely for scripting. JavaScript is a separate topic but many examples in this book include embedded JavaScript code to provide standalone example files that demonstrate the use of HTML elements.

Link Resources

The **<link>** tag that was used in an earlier example to incorporate a style sheet in an HTML document can also be used to incorporate other resources into a document.

This tag may only appear in the head section of a document, but the head section can contain many **<link>** tags. Each **<link>** tag must contain **rel** and **href** attributes, stating the relationship and location of the link resource. It may also include a **type** attribute where appropriate, to hint at the MIME type of the link resource.

Permitted rel (relationship) values				
alternate	author	bookmark	help	icon
license	next	nofollow	noreferrer	prev
search	stylesheet	tag	shortcut icon	

Hot tip

MIME (Multipart Internet Mail Extension) types describe file types – such as **text/html** for HTML files. You can find the list of official MIME types at **https://www.iana.org/ assignments/media-types/media-types. xhtml**

Many of the link types above are intended to help search engines locate resources associated with that HTML document, and the **<link>** tag may also include a **title** attribute to further describe the resource – for example, a version of the page in another language:

```
<link rel="alternate" type="text/html" href="esp.html"
        title="Esta página en Español - This page in Spanish" >
```

In this case, the location of the resource is specified using a relative address that, by default, the browser will seek in the directory in which the HTML document is located. The browser can, however, be made to seek a relative address in a different directory by inserting a **<base>** tag at the start of the document's head section. Its **href** attribute can then specify the absolute directory address – for example, to specify a separate "resources" directory, like this:

```
<base href= "http://localhost/resources/">
```

Beware

When using a **<base>** element it must be placed in the head section before any **<link>** elements.

It is popular to link an icon resource to display in the web browser. This is named exactly as "favicon.ico" and can be placed in the same directory as the HTML document, or in a directory specified by the **<base>** tag. All browsers recognize any other resources in the directory specified by the **<base>** tag.

1 Create a new HTML document that includes meta data, a linked resource, and areas for style rules and script code

```
<!DOCTYPE HTML>
<html lang="en">
<head>
<meta charset="UTF-8">
<meta  name="viewport"
        content="width=device-width, initial-scale=1">
<link rel="shortcut icon" href="favicon.ico">
<title>Document Title</title>
<style>

</style>
</head>
<body>


<script>

</script>
</body>
</html>
```

This template is the basic HTML document that is used in all ensuing examples to create a new HTML document – only the title changes to suit each example.

2 At the beginning of the head section, insert an element to specify a base "resources" directory
`<base href="http://localhost/resources/">`

favicon.html

3 Change the document title to "Favicon Example," then save the HTML document

4 Open an icon editor and create an icon sized 32 x 32 pixels and save your icon alongside the HTML document, or in the "resources" directory, named as "favicon.ico"

favicon.ico 32px x 32px

5 Open the HTML document in your web browser via a web server to see the icon resource appear in the browser

You can force your browser to refresh the favicon by assigning **favicon.ico?v=2** to the link's **href** attribute.

Favicon Example   localhost/favicon.html

27

Summary

- The Web Hypertext Application Technology Working Group (WHATWG) oversees the HTML Living Standard.

- HyperText Transfer Protocol (HTTP) is the common communication standard used by web servers.

- A Uniform Resource Locator (URL) is an absolute web address comprising protocol, domain, and path components.

- A relative address can reference an adjacent file by its name, and may use the ../ syntax to reference a parent directory.

- Web servers send response headers back to the requesting computer and a copy of the requested file, or an error code.

- Each HTML document should have a document type declaration, a head section, and a body section.

- Information about the document itself is contained in the head section, and content is contained in the body section.

- The document's written language is specified to a **lang** attribute in the opening **<html>** root element.

- The document's character-set encoding is specified to a **charset** attribute in a **<meta>** tag within the head section.

- The document's title is specified between **<title> </title>** tags within the head section.

- The free online W3C validator tool can be used to verify that an HTML document is free of errors.

- Metadata describes the document, and a content description can be used by search engines to index the web page.

- The **<style> </style>** tags can be used to embed style sheets within an HTML document.

- The **<script> </script>** tags can be used to include internal and external JavaScript code in an HTML document.

- The **<link>** tag can be used to embed external style sheets and other resources within an HTML document.

2 Structure Web Pages

Proclaim Headings

HTML heading elements are created using **<h1>**, **<h2>**, **<h3>**, **<h4>**, **<h5>**, and **<h6>** tags. These are ranked in importance by their numeric value – where **<h1>** has the greatest importance, and **<h6>** has the least importance. Each heading requires a matching closing tag and should only contain heading text. Typically, the heading's font size and weight will reflect its importance, but headings also serve other purposes.

Heading elements should be used to implicitly convey the document structure by correctly sequencing them – so **<h2>** elements below an **<h1>** element, **<h3>** elements below an **<h2>** element, and so on. This structure helps readers quickly skim through a document by navigating its headings. Search engine spiders may promote documents that have correctly sequenced headings as they can use the headings in their index. They assume headings are likely to describe their content so it is especially useful to include meta keywords from the document's head section in the document's headings.

The **<h1>** element is by far the most important heading, and should ideally appear only once to proclaim the document heading. Often, this can be a succinct version of the document title. Below that, a number of **<h2>** headings can proclaim topical headings for long documents. Each topic might contain individual article headings within **<h3>** elements, followed by paragraph **<p>** elements containing the actual article content.

Beware

Never use heading elements for their font properties as these can be overridden by style sheet rules – always consider headings to represent structure.

heading.html

 1 Create an HTML document (as the template on page 27)

 2 Within the body section, insert a main document heading
<h1>Document Heading</h1>

 3 Next, within the body section, insert a topic heading
<h2>Topic Heading</h2>

 4 Now, within the body section, insert some article headings followed by paragraphs containing the article content
<h3>Article Heading</h3>
<p>Article content...</p>

<h3>Article Heading</h3>
<p>Article content...</p>

5 Finally, add another topic with two more articles

```
<h2>Topic Heading</h2>

<h3>Article Heading</h3>
<p>Article content...</p>

<h3>Article Heading</h3>
<p>Article content...</p>
```

6 Save the HTML document then open it in your web browser to see the headings and document structure

You will discover more about the **<p>** paragraph element on page 46.

Browser window titled "Heading Example", URL: heading.html

Document Heading

Topic Heading

Article Heading

Article content...

Article Heading

Article content...

Topic Heading

Article Heading

Article content...

Article Heading

Article content...

All screenshots throughout this chapter have added (unlisted) style rules to more clearly illustrate the elements described.

Structural outline
```
└ Document Heading
   ├ Topic Heading
   │  ├ Article Heading
   │  └ Article Heading
   │
   └ Topic Heading
      ├ Article Heading
      └ Article Heading
```

The document structure created by the sequenced headings is known as the document "outline". Properly constructed outlines allow a part of the page, such as a single article, to be easily syndicated into another site. The outline for the document above is illustrated alongside the screenshot above.

31

Group Headings

Headings sometimes have a sub-heading or tagline. For example, a document heading could be marked up like this:

```
<h1>American Airlines</h1>
<h2>Doing What We Do Best</h2>
```

Unfortunately, this would strictly require all subsequent headings to be **<h3>** down to **<h6>** – to maintain a correctly sequenced outline. Fortunately, HTML provides a grouping solution with the **<hgroup> </hgroup>** element. This can be used to enclose both the heading and sub-heading, like this:

```
<hgroup>
<h1>American Airlines</h1>
<h2>Doing What We Do Best</h2>
</hgroup>
```

A document may contain multiple **<hgroup>** elements, and each **<hgroup>** element may contain headings **<h1>** down to **<h6>**.

Complete headers may be enclosed in **<header> </header>** tags to include a heading element, or **<hgroup>** element, along with other introductory items – such as a banner, logo, or a table of contents.

You cannot nest
<header> elements one
within another.

1 Create an HTML document (as the template on page 27)

header.html

2 Within the body section, insert a main document heading that includes a banner image
```
<header>

<img src="header-banner.png" width="500"
      height="72" alt="Banner">

<hgroup>
<h1>HTML</h1>
<h2>Building better websites</h2>
</hgroup>

</header>
```

You will discover more
about the **<img>** image
element on page 94.

3 Next, within the body section, insert a topic and article
```
<hgroup>
<h2>Topic Heading</h2>
<h3>Article Heading</h3>
</hgroup>

<p> Article Content...</p>
```

4 Now, within the body section, insert a second topic with a single article
```
<hgroup>
<h1>CSS</h1>
<h2>Cascading Style Sheets</h2>
<h3>Article Heading</h3>
</hgroup>

<p>Article content...</p>
```

5 Save the HTML document then open it in your web browser to see the grouped headings and document header

The **<hgroup>** element only groups headings **<h1>** to **<h6>**, but the **<header>** element groups headings and additional items.

33

Structural outline

- HTML: Building better websites
  - Topic Heading: Article Heading
- CSS: Cascading Style Sheets: Article Heading

Include Navigation

Groups of hyperlinks on a web page, which enable the user to navigate around the page or website, should be enclosed between **<nav> </nav>** tags, or **<menu> </menu>** tags for other links.

This group of links may typically be a horizontal menu in the document header – often called a navigation bar ("nav bar") – or may be a vertical menu down the edge of the page. Note that the **<nav>** element is simply a wrapper around the menu – it does not replace any structural elements.

nav.html

1. Create an HTML document (as the template on page 27)

2. Within the body, insert a header containing a logo, document heading, and horizontal page navigation bar
   ```
   <header>

   <img id="logo" src="nav-logo.png" alt="Logo">

   <h1>Building better websites</h1>

   <nav id="horizontal">
   <h2>Site Links</h2> <p>
   <a href="#html">Markup</a> |
   <a href="#js">Scripting</a> |
   <a href="#css">Style Sheets</a> </p>
   </nav>

   </header>
   ```

Hot tip

You will discover more about the **<a>** anchor element on page 68 and find details about the **
** line break element on page 46.

3. Next, in the body, insert a vertical site navigation menu
   ```
   <nav id="vertical" >
   <h2>Page Links</h2>
   <p>Further Reading<br>In Easy Steps:
   <br><br> <a href="nav-js.html">JavaScript</a>
   <br><br> <a href="nav-css.html">CSS</a> </p>
   </nav>
   ```

4. Now, add topic headings and content
   ```
   <h2 id="html">HTML</h2>
   <p>All about markup...</p>

   <h2 id="js">JavaScript</h2
   <p>All about scripting...</p>

   <h2 id="css" >CSS</h2>
   <p>All about stylesheets...</p>
   ```

Beware

Not every group of hyperlinks is eligible to be contained in a **<nav>** element – only those that provide page-wide or site-wide navigation.

5 Add a style sheet to position the logo image, horizontal navigation bar, and vertical navigation menu

```
<style>
#logo   { float : left ; }
#horizontal      { padding-left : 100px ; display : block ; }
#vertical { float : left ; padding : 0px 30px 30px 30px ; }
</style>
```

6 Save the HTML document, then open the web page in your browser and try out the navigation links

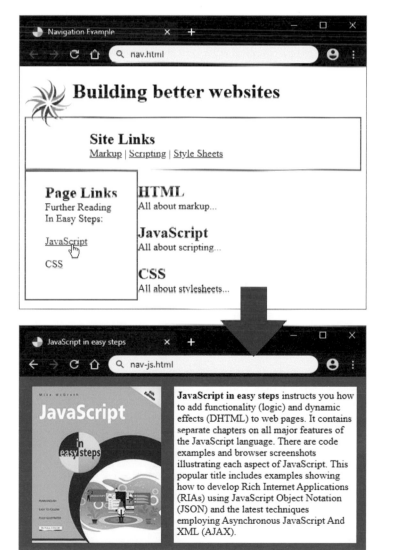

Structural outline
    └ Building better websites
        ├ Site Links
        ├ Page Links
        ├ HTML
        ├ JavaScript
        └ CSS

35

Hot tip

It is popular to create vertical navigation menus as unordered lists – see page 78.

Complete Framework

Just as a typical HTML document may contain a document heading or header group it may also contain a footer, or footer group, at the bottom of the page. The content of each footer is contained between **<footer> </footer>** tags and provides information about that part of the document.

Typically, a **<footer>** element might contain the author's name, the author's contact details within an **<address>** element, or copyright and legal disclaimers within a **<small>** element. Like a **<header>** element, a **<footer>** element can also contain hyperlinks for page and site navigation within a nested **<nav>** element.

The document heading and footer can sensibly be separated by a **<main> </main>** element that will contain the page content.

framework.html

1 Create an HTML document (as the template on page 27)

2 Within the body, insert a document heading
<h1 id="top">Interesting Articles</h1>

3 Next, add a main content container
<main>

<!-- Page content to be inserted here. -->

</main>

4 In the main content container, insert two articles that are the main page content
<article>

<h2 id="art-1">Sally's Article</h2>
<p>Article content...</p>

</article>

<article>

<h2 id="art-2">Terry's Article</h2>
<p>Article content...</p>

</article>

The HTML **<article>** elements might also each contain a **<footer>** element providing contact details for the article's author.

5 Finally, within the body, insert a document footer containing page navigation hyperlinks, copyright details, and a URL address

```
<footer>

<nav>
<h3>Information</h3>
<a href="#art-1">Sally's Article</a> -
<a href="#art-2">Terry's Article</a> -
<a href="#top">Top of Page</a>
</nav>

<small>Copyright &copy;  Example Corporation</small>
<address>www.example.com</address>

</footer>
```

6 Save the HTML document, then open the web page in your browser to see that the document structure comprises a heading, page content, and footer area

Structural outline
└ Interesting Articles
  ├ Sally's Article
  ├ Terry's Article
  └ Information

Create Sections

In HTML all content within the **<body>** element is considered to be part of a "section". Section limits are defined implicitly by correctly sequenced headings in the document outline. Section limits are defined explicitly by placing content within the **<header>**, **<main>**, and **<footer>** framework elements demonstrated in the previous example on pages 36-37.

Page content within the document body or **<main>** element can also be arranged in sections between **<section> </section>** tags. Each section might typically begin with its own heading followed by articles. Similarly, each article might typically begin with its own heading followed by one or more paragraphs.

In understanding the **<section>** and **<article>** elements it helps to consider the way a newspaper contains various sections – news, sport, real estate, and so on. Each section contains various articles.

section.html

 Create an HTML document (as the template on page 27)

 Within the body, insert a document heading
<h1>Newspaper</h1>

Next, add a main content container
<main>

<!-- Page content to be inserted here. -->

</main>

 Now, in the main content container, insert a section containing two articles
<section>
<h2>News Section</h2>

<article>
<h3>Article #1</h3>
<p>Article content...</p>
</article>

<article>
<h3>Article #2</h3>
<p>Article content...</p>
</article>

</section>

<section> elements are not required in short documents like this one – unless you particularly want to add section headings and footers.

5 Next in the main content container, insert another section containing a single article

```
<section>
<h2>Sport Section</h2>

<article>
<h3>Article #1</h3>
<p>Article content...</p>
</article>

</section>
```

6 After the main content container, add a page footer

```
<footer>
<small>Copyright &copy;  Example Corporation</small>
</footer>
```

7 Save the HTML document then open it in your browser to see the article content displayed in sections

Newspaper

News Section

Article #1

Article content...

Article #2

Article content...

Sport Section

Article #1

Article content...

Copyright © Example Corporation

Structural outline
```
└ Newspaper
        ├ News Section
        │       ├ Article #1
        │       └ Article #2
        └ Sport Section
                └ Article #1
```

Provide Asides

HTML usefully provides **<aside> </aside>** tags that can be nested within an **<article>** element in order to incorporate content that is somewhat related to the main content of that article. These allow for supplemental yet separate content to be included – typically displayed as a sidebar or footnote.

Content within an **<aside>** element should be stand-alone information that is related to the article, such as pull-quotes extracted from an affiliated article, a glossary of terms used within the article, or even hyperlinks to pages providing further reading associated with the article.

Alternatively, the **<aside>** element can be used alone, without an **<article>** element, to contain secondary content that is related to the entire page, such as related advertising or a web blog.

 1 Create an HTML document (as the template on page 27)

 2 Within the body, insert a document heading
<h1>Famous Quotes</h1>

aside.html

 3 Next, add a main content container
<main>

<!-- Page content to be inserted here. -->

</main>

 4 Now, in the main content container, insert an article containing a heading, a paragraph, and an aside element
<article>
<h2>Happiness</h2>

**<p><q>The secret of happiness is not in doing what one likes, but in liking what one has to do.</q>
**
<cite>James M. Barrie</cite></p>

<aside>James M. Barrie (1860 - 1937) was a Scottish author and playwright.</aside>

</article>

The HTML **<aside>** and **<nav>** elements may also each contain a **<footer>** element.

...cont'd

5 In the main content container, insert a second, similar article – containing a **class** attribute for sidebar styling

```
<article class="sidebar">
<h2>Cynicism</h2>

<p> <q>A cynic is a man who knows the price of
everything<br>but the value of nothing.</q> <br>
<cite>Oscar Wilde</cite></p>

<aside>Oscar Wilde (1854 – 1900)<br>
was an Irish writer and poet.</aside>

</article>
```

Avoid using the **<aside>** element to contain unrelated advertising.

6 Add a page footer after the main content container, then add a style sheet to control the position of the aside

```
<footer>
<small>Copyright &copy;  Example Corporation</small>
</footer>

<style>
article.sidebar > p,aside
{ display : table-cell ; padding-right : 20px ; }
</style>
```

Do not use the **<aside>** element to contain navigation hyperlinks – those should always be contained inside a **<nav>** element.

7 Save the HTML document and the style sheet, then open the web page in your browser to see how the asides appear

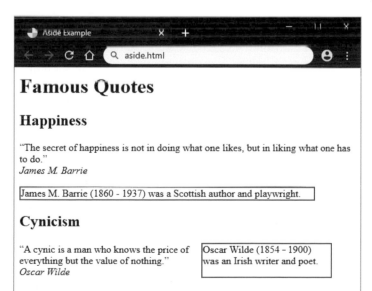

Revise Divisions

The **<div> </div>** division tags, which were used widely in earlier versions of HTML, continue to be supported for backward-compatibility – but the **<div>** element provides no semantic meaning so is best avoided in favor of more meaningful tags.

Unlike other meaningful elements such as **<header>**, **<main>**, **<section>**, **<article>**, **<nav>**, and **<footer>**, the meaningless **<div>** element is anonymous. For example, a smart browser might have a shortcut key to jump to the page's navigation section. This section is easily identifiable when contained in a meaningful **<nav>** element, but not so obvious when contained in a meaningless **<div>** element.

The **<div>** element remains useful for styling purposes, as do the similarly anonymous **<span> </span>** tags. Although the **<div>** and **<span>** elements are meaningless alone, they can include an identifying **id**, **class**, or **style** attribute for application of style rules.

Documents that use the **<div>** element for structural rather than stylistic purposes should be revised to use meaningful elements instead – for example, given the document elements below:

Only use the **<div>** element for styling – always look for a meaningful element to use instead.

divided.html

```
<body>

<div class="header">
<h1>Web Languages</h1>
</div>

<div class="nav">
<h2>Menu</h2>
<p><a href="nav-js.html">JavaScript</a></p>
<p><a href="nav-css.html">Cascading Style Sheets</a></p>
</div>

<div class="main">
<h2> <span style="font-style:italic" > HyperText</span>
Markup Language</h2>
<p>All about HTML...</p>

<h2><span style="font-style:italic" >eXtensible </span>
Markup Language</h2>
<p>All about XML...</p>
</div>

<div class="footer">
<p><small>Copyright &copy; Example Corporation</small></p>
</div>

</body>
```

Structural outline
- Web Languages
  - Menu
  - HyperText Markup Language
  - eXtensible Markup Language

...cont'd

1 Replace the "header" class **\<div>** with a **\<header>** element

2 Replace the "nav" class **\<div>** with a **\<nav>** element

3 Replace the "main" class **\<div>** with a **\<main>** element

4 Add **\<article>** elements around heading and paragraphs, then replace the **\<span>** elements with **\<em>** elements – for automatic emphasis

```
<article>

<h2><em> HyperText</em> Markup Language</h2>
<p>All about HTML...</p>

</article>

<article>

<h2><em>eXtensible</em>Markup Language</h2>
<p>All about XML...</p>

</article>
```

5 Replace the "footer" class **\<div>** with a **\<footer>** element

6 Save the edited document, then open both documents in your browser to see they look identical – the structure is the same but the revision gives semantic meaning

revised.html

You will discover more about the **\<em>** emphasis element on page 50.

Also amend any associated style sheets to select the new elements.

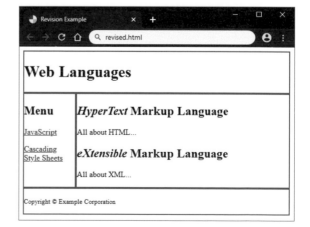

43

Summary

- Heading elements **\<h1>**, **\<h2>**, **\<h3>**, **\<h4>**, **\<h5>**, and **\<h6>** are ranked in order of importance from **\<h1>** down to **\<h6>**.

- Correctly sequenced heading elements implicitly convey the document structure – to create the document outline.

- The **\<hgroup>** element can be used to enclose both a heading and sub-headings – from **\<h1>** down to **\<h6>**.

- Complete headers, including a logo, banner, and headings **\<h1>** to **\<h6>** can be enclosed in a **\<header>** element.

- Groups of hyperlinks providing page or site navigation should be enclosed within a **\<nav>** element.

- A **\<nav>** element is just a wrapper around a menu, typically displayed horizontally in the header or vertically in a sidebar.

- A web page body section framework may comprise **\<header>**, **\<main>**, and **\<footer>** elements.

- Typically, a **\<footer>** element might contain contact details in an **\<address>** element or legal details in a **\<small>** element.

- Each document **\<section>** element will typically begin with a section heading, followed by one or more articles.

- Each document **\<article>** element will typically begin with an article heading, followed by one or more paragraphs.

- Stand-alone information related to an article can be enclosed within an **\<aside>** element nested in an **\<article>** element.

- The **\<div>** and **\<span>** elements are meaningless but are useful for styling purposes.

3 Manage Text Content

This chapter demonstrates how to include text and hyperlinks in page content.

Insert Paragraphs

All text content is traditionally separated into sentences and paragraphs to be more easily read and more readily understood. This is also true for text content in HTML documents, and their paragraphs are contained within **<p> </p>** tags. Each paragraph element is visually separated from the next one by the browser – typically leaving two empty lines between them.

Text within a paragraph will normally automatically wrap to the next line when it meets the element's edge, but it can be forced to wrap sooner by inserting a line break **
** tag.

For emphasis, a horizontal rule **<hr>** tag can be inserted between paragraphs to draw a line separating them. The **<hr>** tag cannot, however, be inserted inside a paragraph to separate sentences. You may be surprised to find the **<hr>** tag in HTML as it would seem to perform a purely presentational function. It is, however, described in the specifications as representing a "paragraph-level thematic break", such as a scene change in a story.

The **
** tag and **<hr>** tag are both single tags that need no matching closing tag.

para.html

1 Create an HTML document

2 Insert a large heading within the body section
<h1>The Statue of Liberty</h1>

3 Next, add a paragraph within the body section
<p>The Statue of Liberty was built over nine years by French sculptor Auguste Bartholdi. Upon its completion in 1884 all 350 individual pieces of the statue were packed into 214 crates for the long boat ride from France to New York.</p>

4 After the paragraph, add a horizontal ruled line
<hr>

5 After the horizontal ruled line, add a second paragraph
<p>The statue arrived in America several months later and was reconstructed on Liberty Island. Auguste Bartholdi thought that the New York harbor was the perfect setting for his masterpiece because it was where immigrants got their first view of the New World.</p>

6 Now, insert breaks into the paragraphs to control the length of their lines

```
<p>The Statue of Liberty was built over nine years
<br>by French sculptor Auguste Bartholdi.
<br>Upon its completion in 1884 all 350 individual pieces
of the statue were packed into 214 crates for the long
boat ride from France to New York.</p>

<p>The statue arrived in America several months later
<br>and was reconstructed on Liberty Island.
<br>Auguste Bartholdi thought that the New York harbor
was the perfect setting for his masterpiece because it was
where immigrants got their first view of the New World.
</p>
```

Hot tip

The **<hr>** element can be considered to be the HTML equivalent of the *** section separator often found in stories and essays.

7 Save the HTML document then open it in your web browser to see the heading, paragraphs, forced line breaks, and horizontal ruled line

---

Paragraph Example ✕ +

← → C ⌂ 🔍 para.html 🔘 ⋮

The Statue of Liberty

The Statue of Liberty was built over nine years
by French sculptor Auguste Bartholdi.
Upon its completion in 1884 all 350 individual pieces of the statue were packed into 214 crates for the long boat ride from France to New York.

The statue arrived in America several months later
and was reconstructed on Liberty Island.
Auguste Bartholdi thought that the New York harbor was the perfect setting for his masterpiece because it was where immigrants got their first view of the New World.

Include Quotations

It is important to recognize that some HTML elements produce a rectangular block area on the page in which to display content, while others merely produce a small block on a line within an outer containing block. These are referred to as "flow" and "phrasing" elements. Phrasing elements, which produce a small block on a line, must always be enclosed by a flow element, which produces the larger containing block, such as **<p> </p>**. The difference between flow elements and phrasing elements can be seen by contrasting how web browsers display the two HTML elements that are used to include quotations in documents.

The **<blockquote> </blockquote>** tags are intended to surround long quotations from another source, which can be specified by its **cite** attribute. For this element, the browser typically produces a rectangular block area to contain the quotation, starting on a new line and indented from surrounding content – so **<blockquote>** is a flow element.

The **<q> </q>** tags, on the other hand, are intended to surround short quotations from another source, which can be specified by its **cite** attribute. For this element, the browser typically produces a small block area on the current line to contain the quotation – so **<q>** is a phrasing element.

Unlike the **<blockquote>** flow element, the **<q>** phrasing element causes the browser to automatically add quotation marks around the element's content when it gets displayed on the page. Ideally, these should be double quotation marks surrounding the entire element content, and single quotation marks around any inner nested quotations, but its implementation may vary.

quote.html

1 Create an HTML document

2 Insert a paragraph within the body section
<p>A Paragraph Flow Block!</p>

3 Within the body section, insert a blockquote containing two small nested quotations
<blockquote cite="http://www.example.com/origin.html">
**A Blockquote Flow Block!
Paul said, <q>I saw Emma**
at lunch, she told me <q>Susan wants you to get some
ice cream on your way home.</q> I think I will get some
at Ben and Jerry's on Main Street.</q> </blockquote>

4 Save the HTML document then open it to compare the double quote marks, single quote marks, and apostrophe

Quotation Example   × +   — □ ×

← → C ⌂   🔍 quote.html   👤 ⋮

A Paragraph Flow Block!

> A Blockquote Flow Block!
> Paul said, "I saw Emma at lunch, she told me 'Susan wants you to get some ice cream on your way home.' I think I will get some at Ben and Jerry's on Main Street."

Hot tip

By default, the paragraph element block will fill the width of its containing element – like the **<h1>** element block in the example on page 35.

5 Insert this style sheet into the head section of the document, then reload the page to reveal the blocks

```
<style>
p, blockquote { border : 2px solid Green ; }
q { background : LawnGreen ; }
</style>
```

Quotation Example   × +   — □ X

← → C ⌂   🔍 quote.html   👤 ⋮

A Paragraph Flow Block!

> A Blockquote Flow Block!
> Paul said, "I saw Emma at lunch, she told me 'Susan wants you to get some ice cream on your way home.' I think I will get some at Ben and Jerry's on Main Street."

Don't forget

HTML recognizes the standard color names described at **w3.org/TR/css-color-3/#svg-color** The names are not case-sensitive, but can usefully be capitalized – for example, **LawnGreen** is more readable than **lawngreen**. The flow elements are shown here with **Green** solid borders, and the phrasing elements with a **LawnGreen** background.

Add Emphasis

HTML provides four phrasing elements that can be used to emphasize text within the body of a document:

- Text enclosed between **<b> </b>** tags is enhanced without conveying extra importance, such as keywords in a paragraph – typically displayed in a bold font.

- Text enclosed between **<i> </i>** tags is enhanced without conveying extra importance, such as technical terms in a paragraph – typically displayed in an italic font.

- Text enclosed between **<strong> </strong>** tags gains increased importance, without changing the meaning of the sentence – typically displayed in a bold font.

- Text enclosed between **<em> </em>** tags should be stressed to deliberately affect the meaning of the sentence – typically displayed in an italic font.

It is perhaps surprising that the **<b>** and **<i>** tags remain in HTML, as they outwardly suggest that content should be presented in a bold or italic font – contradicting the aim of HTML to separate structure from presentation. According to the specifications, their meaning has been redefined, however, so content within a **<b>** element should be "stylistically offset" and that within an **<i>** element should be seen as in an "alternate voice". In real terms, these are nonetheless represented by bold and italic fonts, but should only be used as a last resort as they do not convey meaning – use **<strong>** and **<em>** tags instead.

The advantage of the **<strong>** and **<em>** tags is that they describe the importance of their content relative to surrounding text and let the browser choose how it should be presented. Additionally, these tags are more relevant to suggest how narrators should convey their content vocally.

As with many HTML tags, the **<strong>** and **<em>** tags can be nested but care must be taken to close nested elements correctly. For example, **<strong><em>...</em></strong>** is the correct order, whereas **<strong><em>...</strong></em>** is incorrect and will not validate.

Don't forget

The specifications encourage web page authors to consider accessibility issues in all aspects of their web page designs.

1 Create an HTML document

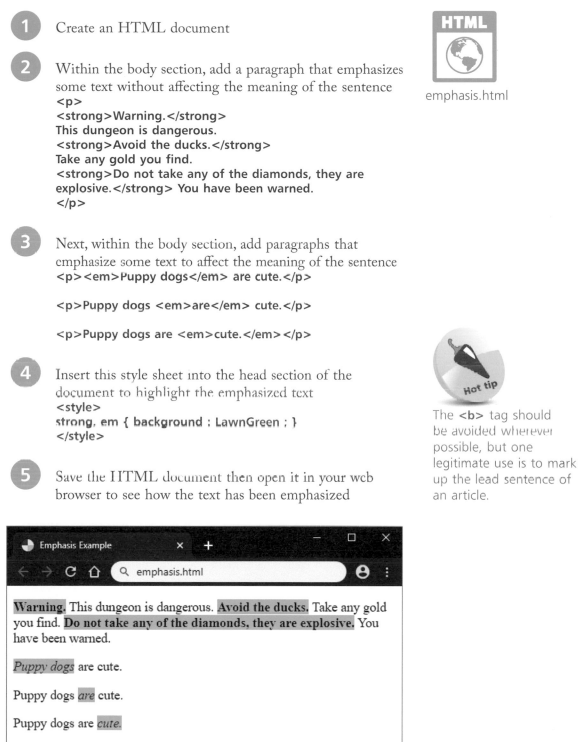

emphasis.html

2 Within the body section, add a paragraph that emphasizes some text without affecting the meaning of the sentence
```
<p>
<strong>Warning.</strong>
This dungeon is dangerous.
<strong>Avoid the ducks.</strong>
Take any gold you find.
<strong>Do not take any of the diamonds, they are
explosive.</strong> You have been warned.
</p>
```

3 Next, within the body section, add paragraphs that emphasize some text to affect the meaning of the sentence
```
<p><em>Puppy dogs</em> are cute.</p>

<p>Puppy dogs <em>are</em> cute.</p>

<p>Puppy dogs are <em>cute.</em></p>
```

4 Insert this style sheet into the head section of the document to highlight the emphasized text
```
<style>
strong, em { background : LawnGreen ; }
</style>
```

5 Save the HTML document then open it in your web browser to see how the text has been emphasized

Hot tip

The **<b>** tag should be avoided wherever possible, but one legitimate use is to mark up the lead sentence of an article.

51

Add Modifications

HTML provides three elements that can be used to format text within the body of a document:

- Text enclosed between **\<small\> \</small\>** tags is regarded as a side comment to surrounding text, such as copyright information – typically displayed in a smaller font.

- Text enclosed within **\<del\> \</del\>** tags is regarded as having been removed from the document, such as a completed item in a to-do list – typically displayed with a strike-through line.

- Text enclosed within **\<ins\> \</ins\>** tags is regarded as having been added to the document, such as a new additional item in a "to do" list – typically displayed with an underline.

The **\<small\>** tag is only meant to contain short comments that supplement surrounding content. It is not intended for use with large sections of text, such as multiple paragraphs, as that would be considerably more than a side comment.

In displaying content contained within a **\<small\>** element, the web browser considers the size of the font used to display the surrounding content, then applies an appropriate reduction. Therefore, where the surrounding content is displayed with a font of 12-point size, content contained within a **\<small\>** element might be displayed with a font of 10-point size – the precise size is determined by the browser.

Both **\<del\>** and **\<ins\>** elements can be used within a section of content, to markup snippets of changed text, and to enclose entire sections of changed content, such as replaced paragraphs.

The **\<del\>** and **\<ins\>** tags may optionally include a **cite** attribute to specify the URL of a document explaining the changes made.

format.html

1. Create an HTML document

2. Within the body section, insert a paragraph containing a side comment for legal purposes
   ```
   <p>Example Corp today announced record profits for
   the second quarter <small>(Full Disclosure: EG News
   is a subsidiary of Example Corp)</small>, leading to
   speculation about a merger with Demo Group.</p>
   ```

3 Next, insert a large heading and a regular paragraph
```
<h1>To Do List</h1>
<p>Empty the dishwasher</p>
```

4 Now, insert a paragraph that has been deleted
```
<del><p>Take out the trash</p></del>
```

5 Then, insert a paragraph that has been added
```
<ins><p>Sweep the yard</p></ins>
```

6 Finally, insert a paragraph that has been added, which contains a text snippet that has been changed
```
<ins>
<p>Feed the <del>dog</del><ins> cat</ins></p>
</ins>
```

7 Save the HTML document then open it in your web browser to see how the text has been formatted

---

Format Example   ✕   ✛   — ◻ ✕

← → C ⌂   🔍 format.html   ⊖ ⋮

Example Corp today announced record profits for the second quarter (Full Disclosure: EG News is a subsidiary of Example Corp), leading to speculation about a merger with Demo Group.

To Do List

Empty the dishwasher

~~Take out the trash~~

<u>Sweep the yard</u>

<u>Feed the ~~dog~~ cat</u>

The **\<small\>** tag does not denote content of lesser importance, only that it is a side comment to surrounding text.

53

Add Phrasing

HTML provides four phrasing elements that can be used to mark text for special treatment within the body of a document:

● Text enclosed between **<s> </s>** tags is marked as being superseded by more accurate or relevant up-to-date content – typically displayed with a strike-through line.

● Text enclosed between **<u> </u>** tags is marked as being different in some way to normal text content – typically displayed with an underscore line to underline the text.

● Text enclosed between **<mark> </mark>** is marked as being of special significance for reference – typically displayed in a colored background block to highlight the text.

● Text broken by a **<wbr>** tag is invisibly marked as being a suitable point at which to break a line of text – representing a word-break opportunity.

It is important to note that specifications state that the **<s>** tag should not be used to indicate edited content within a document. The **<del>** tag should be used instead to indicate document edits.

Similarly, the **<mark>** tag should not be used to emphasize the importance of text content, but should only be used to highlight the relevance of text within a document. The **<strong>** and **<em>** tags should be used instead to indicate emphasis.

The **<u>** tag was deprecated in the HTML5.0 specification, as underlined text within a document traditionally indicates hyperlinks. The **<u>** tag has, however, been reinstated for the purposes of labeling misspelled words or proper names in Chinese. Authors are nonetheless strongly discouraged from using the **<u>** tag for emphasis, to avoid confusion with hyperlinks. Once again, the **<strong>** and **<em>** tags should be used instead to indicate emphasis.

Where the document contains lengthy content that may exceed the width of the browser, you may wish to use the **<wbr>** tag to indicate appropriate points at which a line-break can be inserted.

Use style sheet rules for presentation purposes rather than the **<u>** tag for underlines.

1 Create an HTML document

mark.html

2 Within the body section, add a paragraph that marks a word-break opportunity
```
<p>Microsoft Surface Pro 7
<wbr>- 256GB / Intel Core i7</p>
```

3 Next, within the body section, add a paragraph that marks a superseded price and provides a current price
```
<p><s>$1,499</s> $1,299</p>
```

4 Now, within the body section, add paragraphs that mark text for reference and mark a misspelled word
```
<p>Memory: <mark>16GB</mark>
<br>Screen: <mark>12.3-inch</mark></p>
<p>Surface <u>Penn</u> Included</p>
```

5 Save the HTML document then open it in your web browser to see how the text has been marked

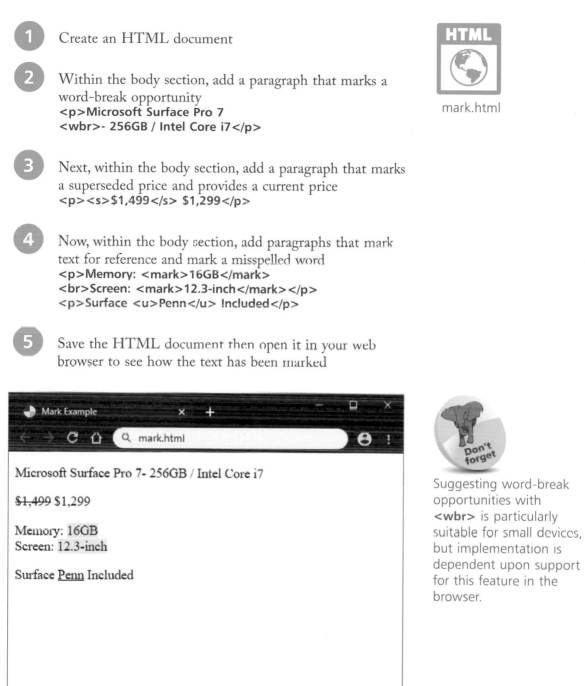

Suggesting word-break opportunities with **<wbr>** is particularly suitable for small devices, but implementation is dependent upon support for this feature in the browser.

Retain Formatting

Where it is desirable to have the browser render text content that has been "preformatted", the web page author can enclose that content between **<pre> </pre>** flow element tags. These advise the browser that the following instructions should be applied:

● Preserve white space.

● Render all text with a fixed-width font.

● Disable automatic word-wrapping.

● Do not disable bi-directional processing.

Preserving the white space retains all spaces, tabs, and line breaks. This is great to display lengthy poems in which every second line is indented – for example, with this verse:

```
ReadingGaol.txt - Notepad                    —   □   ×
File  Edit  Format  View  Help
In Debtors' Yard the stones are hard,
    And the dripping wall is high,
So it was there he took the air
    Beneath the leaden sky,
And by each side a Warder walked,
    For fear the man might die.
```

In this case, each second line is indented by four character widths – created by hitting the space bar four times to insert four invisible space characters. These indents will be exactly preserved by the **<pre>** element as four character widths.

Tab characters, on the other hand, can present some surprises as they are usually interpreted by a browser as eight character widths. This agrees with the tab size in Windows' Notepad application but other text editors can vary. This means that preformatted text containing tab characters may appear to be misaligned by the **<pre>** element. It is for this reason that the specifications discourage the use of tab characters when creating preformatted text content.

The **<pre> </pre>** tags can also be useful to ensure "Text-Art" (sometimes used as web forum signatures) will appear as intended.

Beware

Use spaces rather than tabs when preparing preformatted text.

preformat.html

1 Create an HTML document

2 Within the body section, insert a document heading
<h1>Text-Art Signature</h1>

3 Ensure that the font in your text editor is set to a fixed
width font, such as Lucida Console for Notepad

4 Next, in the body section, insert a **<pre>** element
containing preformatted content in a fixed width font –
and produced without any tab characters

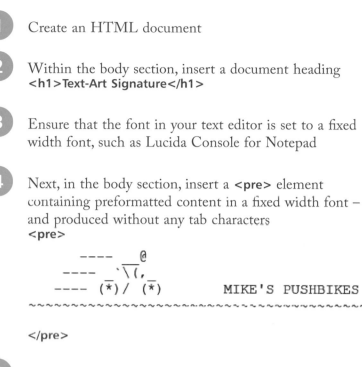

Notice that **<pre>** is a
flow element so it does
not need to be enclosed
within a paragraph – it
creates its own block.

5 Save the HTML document then open it in your web
browser to ensure the content retains preformatting

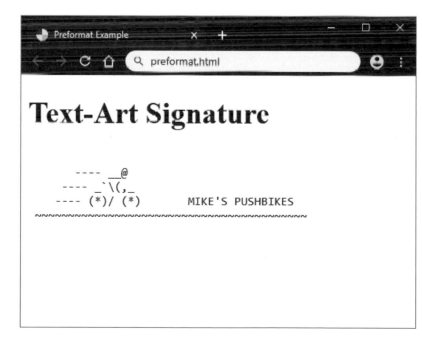

You can use any
character within a fixed
width font to create
your Text-Art – Windows
users can use the
Character Map program
in System Tools to select
special characters from
the Lucida Console font.

Use Superscript

Regular text in a paragraph area of a web page is displayed in invisible inline phrasing boxes that comprise an outer logical box, and an inner font box containing a baseline:

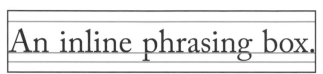

An inline phrasing box.

Logical box
Baseline
Font box

The vertical line spacing is determined by the font height to allow space between characters that extend below the baseline, such as "p", and tall characters that extend upwards, such as "b", plus a vertical margin area.

> Text in a paragraph written in an inline phrasing box.
> Lines are spaced so characters do not collide. ™

You can find a chart of all character entities at dev.w3.org/html5/html-author/charref

Additionally, the font box will accommodate "superscript", such as the trade mark symbol ™ produced by the **™** character entity. Superscript is any text, number, or symbol that appears smaller than regular text and is set above the baseline. Mathematical formulae can use superscript to indicate numeric powers with the character entities **²** for ^{2} and **³** for ^{3}. The font box will also accommodate "subscript" – that appears smaller than regular text and is set below the baseline.

The height available for superscript and subscript with the standard vertical line spacing is limited so the character size is restricted. Rather than use character entities for this purpose it is often better to use the HTML **\[\]** tags for superscript and **_** tags for subscript. These elements increase the vertical line spacing to allow more prominent superscript and subscript characters. For example, **\^{2\}** is larger than **²**. Additionally, any content can be included within these elements so you are not restricted to available character entity references.

> Lines are spaced so characters do not collide with the superscript below.
> Text line in a paragraph containing ^{superscript} and $_{subscript}$
> Lines are spaced so characters do not collide with the subscript above.

1 Create an HTML document

modify.html

2 Within the body section, insert a paragraph containing superscript produced by character entities
```
<p>
Square of four: 4&sup2; = 16 <br>
Cube of four: 4&sup3; = 64
</p>
```

3 Now, in the body section, insert a similar paragraph containing superscript produced by HTML elements
```
<p>
Square of four: 4<sup>2</sup> = 16 <br>
Cube of four: 4<sup>3</sup> = 64
</p>
```

4 Finally, in the body section, insert a paragraph containing subscript produced by HTML elements
```
<p>
Water: H<sub>2</sub>O <br>
Oil of Vitriol: H<sub>2</sub>SO<sub>4</sub>
</p>
```

5 Save the document then open it in your browser to compare the superscript and to see the subscript text

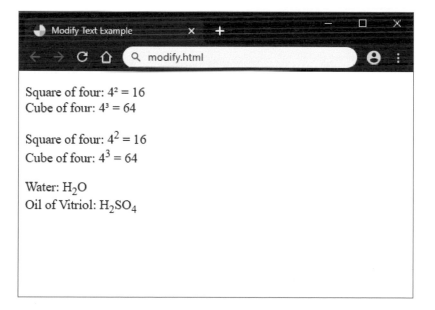

Hot tip

When using superscript ^{2} in paragraphs to denote area, such as 10 feet^2, you may prefer to use the entity **²** rather than **²** to keep line spacings equal.

Display Code

HTML provides five phrasing elements specifically to include computer program code within the body of a document:

- Complete program code, or snippets, can be enclosed between **<code> </code>** tags for display in a suitable font.

- Program variable instances can be enclosed between **<var> </var>** tags to differentiate them from regular text.

- Sample program input and output can be enclosed between **<samp> </samp>** tags to differentiate them from regular text.

- Content that also has associated machine-readable code can be enclosed between **<data> </data>** tags and the code specified to its required **value** attribute.

- Dates and times can be enclosed in **<time> </time>** tags and a machine-readable version specified to its **datetime** attribute.

The **<data>** element could, for example, describe a book title and its machine-readable ISBN, then the **<time>** element could describe that book's publication date:

code.html

The datetime value of a **<time>** element must be in a valid format – for example, as full datetime with **2023-12-25 14:30** or month as **2023-12** or date as **2023-12-25** or day with **12-25** or time only as **14:30**.

1. Create an HTML document

2. In the body section, insert a program description containing variables, sample input, and sample output

```
<p>
This program assigns an input value to
<var>degF</var>
then performs a conversion on that value, assigning the
result to
<var>degC</var>
for output. For example, input of
<samp>98.6</samp>
will output
<samp>37C</samp>.
</p>
```

3. Now, in the body section, state the program code source

```
<data value="978-1-84078-719-1">
C++ Programming in easy steps, 5th Edition</data>
<time datetime="2023-12-15">
(December 25th, 2023)</time>
```

4 Next, in the body section, insert the preformatted program code

```
<pre>
<code>
#include &lt;iostream&gt;
using namespace std;

int main()
{
  float degF, degC;
  cout &lt;&lt; "Enter Fahrenheit Temperature: ";
  cin &gt;&gt; degF;
  degC = ((degF - 32.0 ) * (5.0 / 9.0));
  cout &lt;&lt; degF &lt;&lt; "F is " &lt;&lt; degC &lt;&lt; "C";
  cout &lt;&lt; endl;
  return 0;
}
</code>
</pre>
```

Note that all angled bracket characters in the program code have been replaced by character entities to avoid conflict with the HTML tags.

5 Save the HTML document then open it in your web browser to see how the program description, source details, and program code is displayed

61

Program Code Example

code.html

This program assigns an input value to *degF* then performs a conversion on that value, assigning the result to *degC* for output. For example, input of 98.6 will output 37C.

C++ Programming in easy steps, 5th Edition (December 25th, 2023)

```
#include <iostream>
using namespace std;

int main()
{
  float degF, degC;
  cout << "Enter Fahrenheit Temperature: ";
  cin >> degF;
  degC = ((degF - 32.0 ) * (5.0 / 9.0));
  cout << degF << "F is " << degC << "C";
  cout << endl;
  return 0;
}
```

Remember to insert the phrasing **<code>** element within a **<pre>** flow element to preserve the program code layout in an HTML document.

Give Advice

HTML provides four phrasing elements that can be used to designate advisory phrases within the body of a document:

- Text can be enclosed between **<abbr> </abbr>** tags to indicate it is an abbreviation.

- Text can be enclosed between **<cite> </cite>** tags to indicate it is a citation or reference from another source.

- Text can be enclosed between **<dfn> </dfn>** tags to indicate it is the definitive instance of that term.

- Text can be enclosed between **<kbd> </kbd>** tags to indicate input to be entered by the user from the keyboard.

Every HTML element that can legally appear within the body of a document may optionally include a **title** attribute. Values specified to a **title** attribute are typically displayed as a tooltip that pops up when the user places the cursor over the element. This means that each of the phrasing elements listed above can include a **title** attribute to expand on the meaning of its content.

advice.html

1 Create an HTML document

2 In the body section, insert a paragraph containing an abbreviation with tooltip advice
```
<p>
<abbr title="HyperText Markup Language">
HTML
</abbr> in easy steps</p>
```

3 Next, insert a citation reference with tooltip advice
```
<p>
<cite title="Inventor of the HyperText Markup
Language">Sir Tim Berners-Lee</cite></p>
```

4 Now, insert a definitive term with tooltip advice
```
<p>
<dfn title="The popular language of the World Wide Web.
Commonly abbreviated to 'HTML'">
HyperText Markup Language</dfn></p>
```

Don't forget

Remember to use single quote marks for nested quotes – as with 'HTML' in Step 4.

5 Then, insert a keyboard instruction with tooltip advice

```
<p>
<kbd title="Press the Y key on your keyboard to execute
a script. This requires JavaScript to be enabled in your
browser">Hit Y to Continue.</kbd></p>
```

6 Finally, add a script that will respond to the keyboard instruction

```
<script>

function showkey( e ) {

  if( e.keyCode === 89 || e.keyCode === 121 )
  {
    alert( 'Y pressed. Thank You.' )
  }
}

document.onkeydown = showkey

</script>
```

Hot tip

The script looks at the keycode when the key gets pressed and will respond to lowercase "y" and uppercase "Y".

7 Save the HTML document, then open it in your browser and place the cursor over the elements to see the tooltips

```
Advice Example        ×   +                              □   ×
←  →  C  ⌂    Q  advice.html                          ⊖   ⋮

HTML in easy steps

Sir Tim Berners-Lee

HyperText Markup Language
                      The popular language of the World Wide Web. Commonly abbreviated to 'HTML'
Hit Y to Continue.
```

8 With JavaScript enabled in your browser, hit the Y key to see the script response

```
This page says

Y pressed. Thank You.

                              OK
```

Gauge Quantity

The value within a range can be represented visually on a web page using an HTML **<meter>** element to display a gauge.

The **<meter>** element must include a **value** attribute that defines a fractional measurement. Optionally, the **<meter>** tag may also include **min** and **max** attributes to specify minimum and maximum range boundaries. If these are omitted, the default range 0-1 is assumed.

The **<meter>** tag may also include **low** and **high** attributes to specify low and high positions within a range, and an **optimum** attribute can specify an ideal preferred position within a range.

The **low** and **high** attributes can specify a sub-optimal range within the overall range specified by the **min** and **max** attributes. This can, in effect, separate the gauge into three parts – low, medium, and high. Although not included in the HTML standard specifications, these three parts can be denoted by the web browser using different colors. For example, the Google Chrome browser sensibly uses red for the low part, yellow for the medium part, and green for the high part.

It is recommended that the **<meter>** element should include text describing the state of the gauge that will be displayed only in browsers that do not display this element visually.

Interactive **<details>** and **<summary>** elements can respond to user actions without scripting to disclose additional information. Typically, the **<details>** element provides a "disclosure widget" on the web page represented by a triangular arrow. A nested **<summary>** element displays a caption describing information hidden within the widget. The additional information is contained within elements nested within the **<details>** element, after the **<summary>** element.

When the user clicks the **<summary>** element, the widget state changes from "closed" to "open" and the hidden information is revealed. Clicking the **<summary>** element once more will close the widget and hide the information again.

The triangular arrow twists around to represent the open and closed state of the widget – consequently, these widgets are sometimes called "twisties".

1 Create an HTML document

meter.html

2 In the body section, insert an article containing a meter with a range 0-100, a sub-optimal range 15-50, an optimum value of 100, and a current value of 80%

```
<article><h2>Gauge</h2>Fuel Level:
<meter min="0" low="15" high="50" max="100"
optimum="100" value="80" >80%</meter>
<!-- Details to be inserted here. -->
</article>
```

3 Insert details describing the current status of the gauge

```
<details>
<summary>Status</summary>
OK to Continue...
</details>
```

4 Repeat Steps 2 and 3 twice, to create two more gauges, then edit their current values to 40 and 10 respectively, and supply appropriate descriptions

5 Save the HTML document then open it in your browser and click each **<summary>** element to see its description

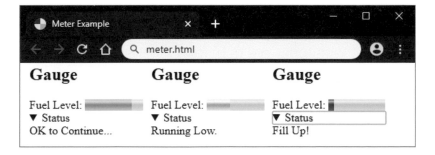

The **<meter>** element should only be used to indicate a fractional measurement within a specified range, not to indicate progress – use the **<progress>** element for that (see page 112).

Without the bi-direction override, these character entities get displayed in their logical order as לארשי – which is back-to-front for the right-to-left Hebrew language.

Direct Language

The recommended UTF-8 document encoding format provides support for bi-directional text, so that characters from languages written right-to-left, such as Hebrew, are automatically written in that direction and may appear alongside left-to-right text such as English. Content to be read in right-to-left direction should be enclosed within **<bdi> </bdi>** bi-directional isolation tags so as not to confuse the browser, as it expects to read left-to-right. Additionally, HTML provides a **<bdo>** bi-direction override element to which a text direction can be explicitly specified as either "ltr" or "rtl" by its **dir** attribute. The bi-direction override allows characters from right-to-left languages to be written as character entities in an HTML document in "logical" left-to-right order, but to be displayed in "visual" right-to-left order. For example, the **<bdo>** element below encloses five character entities from left-to-right, in the order they may have been entered, but displays them right-to-left:

<bdo dir="rtl">ישראל</bdo>

.... appears as ישראל (Yiśrā'ēl in the Latin alphabet).

Ruby annotation

For Eastern languages, HTML supports "Ruby annotation" that usefully provides pronunciation alongside text. In Japanese, for example, there is more than one alphabet. Text written in the semantic "Kanji" alphabet, which has thousands of characters, is often annotated with its equivalent in the phonetic "Hiragana" language, which has around 50 characters, to aid pronunciation. This is called "Furigana" in Japanese and "Ruby" in English – named after the small font used to indicate the pronunciation. For the benefit of Westerners, the Japanese kanji text can be annotated with "Romaji" – its Latin alphabet equivalent. Similarly in Chinese, text written in the "Mandarin" alphabet can be annotated with "Pinyin" – its Latin alphabet equivalent.

HTML Ruby annotation is entirely enclosed between root **<ruby> </ruby>** tags. This element may then enclose the Eastern text within **<rb> </rb>** tags (ruby base) and the pronunciation between **<rt> </rt>** (ruby text) tags. Optionally, an English language equivalent may be provided within **<rtc> </rtc>** tags.

1 Create an HTML document type

ruby.html

2 In the body section, insert an element for Japanese text
with its appropriate pronunciation annotation
<ruby>

```
<!-- Japanese Kanji text. -->
<rb>東京</rb>
<!-- Romaji annotation. -->
<rt>tō kyō</rt>
<!-- English equivalent text. -->
<rtc>Tokyo</rtc>

</ruby>
```

3 Next, insert an element for Chinese text
with its appropriate pronunciation annotation
<ruby>

```
<!-- Chinese Mandarin text. -->
<rb>北京</rb>
<!-- Pinyin annotation. -->
<rt>běi jīng</rt>
<!-- English equivalent text. -->
<rtc>Beijing</rtc>

</ruby>
```

4 Save the HTML document then open it in your browser
to see the text and ruby annotations

Don't confuse ruby annotation with the unconnected Ruby programming language.

Create Hyperlinks

When the internet carried only text content, "hypertext" provided the ability to easily access related documents and was fundamental to the creation of the World Wide Web. Today, images can also be used for this purpose so any navigational element of a web page is now referred to as a "hyperlink" (or simply as a "link").

Hyperlinks are enclosed between **<a>** **</a>** anchor tags, which specify the target URL to an **href** (hypertext reference) attribute in the opening tag. Optionally, the **<a>** tag can also include a **title** attribute to specify text to display in a "tooltip" that will appear when the user places the cursor over the hyperlink.

The web browser will display a hyperlink in a manner that distinguishes it from regular text – typically, hypertext gains an underline and image-based hyperlinks gain a colored border.

Each web page hyperlink is sensitive to three interactive states:

- **Hover** – gaining focus, the cursor is placed over the hyperlink.

- **Active** – retrieving the linked resource, the user clicks the hyperlink.

- **Visited** – the linked resource has previously been retrieved.

Style rules can be used to emphasize each hyperlink state:

link.html

1 Create an HTML document

2 Next, add a link to a style sheet in the head section
<link rel="stylesheet" href="link.css">

3 Now, in the body section, insert a hyperlink to a target page – including tooltip advice
<a href="link-target.html"
title="A hyperlink to a target page">Visit Target</a>

link-target.html

4 Save the HTML document, then create a similar second document that links the same style sheet and contains a hyperlink targeting the first document
<a href="link.html"
title="A hyperlink to return">Return</a>

...cont'd

5 Save the second HTML document alongside the first, then create a style sheet to emphasize each hyperlink state

```
a:hover   { background : Yellow ; }
a:active  { background : Red ; color : White ; }
a:visited { background : Aqua ;  }
```

6 Save the style sheet alongside the two HTML documents, then open the first page in your browser to see the hyperlink in its default state

7 Place the cursor over the hyperlink to see the hyperlink in its hover state

8 Hold down the left mouse button on the hyperlink to see the hyperlink in its active state

9 Release the mouse button to load the target page

10 Click the hyperlink in the target page to reload the first page

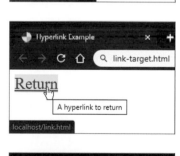

If you click the link to load the target page again, the browser recognizes it has been previously visited from the first page.

11 See that the hyperlink on the first page is now in its visited state

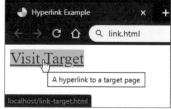

Access Keys

There are three ways to access the target of a hyperlink:

- **Pointer** – a mouse or similar device places a screen pointer over a hyperlink, then the user clicks to access its target.

- **Tab** – repeatedly hit the Tab key to successively focus on each hyperlink in turn, then hit Return to access the target of the currently selected hyperlink.

- **Access Key** – hit a designated character key to focus on a particular hyperlink, then hit Return to access its target.

A designated character key is specified for a hyperlink by the **accesskey** attribute of an **<a>** anchor tag. The method to utilize the designated key generally requires the user to press **Alt** + *accesskey* with most web browsers, such as Google Chrome, but it's **Alt** + **Shift** + *accesskey* with Firefox.

keys-home.html

1 Create an HTML document

2 Next, add a style sheet in the head section to remove default hyperlink styles
```
<style>
a        { text-decoration : none ; color : Black ; }
a:focus { background : Red ; color : White ; }
</style>
```

3 Now, in the body section, insert two hyperlinks that designate different numeric access key characters
```
<a href="keys-home.html" accesskey="1">
Home Page</a> |
<a href="keys-catalog.html" accesskey="2">
Catalog Page</a>
```

keys-catalog.html

4 Save the HTML document then create a similar second document containing the same two hyperlinks – but without the style sheet that removes default styles
```
<a href="keys-home.html" accesskey="1">
Home Page</a> |
<a href="keys-catalog.html" accesskey="2">
Catalog Page</a>
```

5 Save the second HTML document alongside the first, then open the first web page in your browser to see the hyperlinks without their default styles

Removing the default hyperlink styles means they are no longer easily recognizable as links – so it is best avoided unless some other indication makes the user aware they can be used for navigation purposes.

6 Hit the **Tab** key repeatedly until the second hyperlink receives focus, then hit **Return** to follow that link

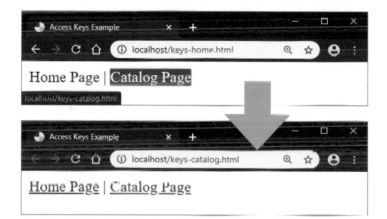

7 Press the access key combination and number 1 key (e.g. **Alt + 1**) then hit **Return** to follow the first hyperlink

Mac users should press **CMD** + *accesskey* with their Safari browser.

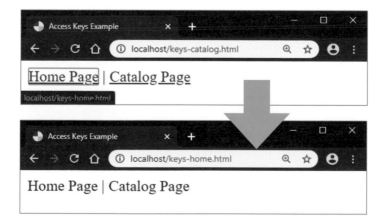

Fragment Links

Hyperlinks can target a specific point in a document that has been created with a "fragment" identifier – an element with a unique identifying name assigned to an **id** attribute in its opening tag. Within the hyperlink, the fragment identifier is specified to an **href** attribute in the opening **<a>** tag prefixed by a # hash character. For example, the tag **<a href="#top">** targets an element within the same document that contains the unique fragment identifier name of "top".

A hyperlink can also target a specific point in a different document using the document's URL, followed by a # hash character, then the fragment identifier. For example, the tag **<a href="index.html#top">** targets an element within a document named "index.html" that contains the unique fragment identifier name of "top".

Following a hyperlink to a fragment identifier displays the document from the point where the fragment identifier appears:

frag.html

 Create an HTML document

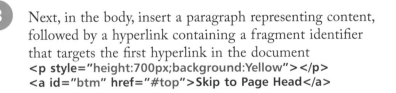 Within the body section, insert two hyperlinks that contain fragment identifiers and also target different fragments
```
<a id="top" href="#btm">Skip to Page Foot</a> |
<a id="next-btm" href="frag-next.html#btm">
Skip to Next Page Foot</a>
```

3 Next, in the body, insert a paragraph representing content, followed by a hyperlink containing a fragment identifier that targets the first hyperlink in the document
```
<p style="height:700px;background:Yellow"></p>
<a id="btm" href="#top">Skip to Page Head</a>
```

frag-next.html

4 Save the HTML document then create a second similar document with hyperlinks both above and below content
```
<a id="top" href="#btm">Skip to Page Foot</a>
<p style="height:700px;background:Red">Content...</p>

<a id="btm" href="#top">Skip to Page Head</a> |
<a id="prev-top" href="frag.html#top">
Skip to Previous Page Head</a>
```

5 Save the second HTML document alongside the first, then open the first page in your browser and click the first hyperlink to go to the bottom of this page

The # hash character is used in HTML to target fragments and to specify hexadecimal color values, and in CSS to select elements by their **id** attribute for styling. Hexadecimal color values specify Red, Green, and Blue components that make up the color – for example, the color Red is hexadecimal #FF0000.

6 Next, click the hyperlink to return to the top of this page

7 Now, click the second hyperlink to go to the bottom of the next page

At the end of lengthy pages include a hyperlink to a fragment at the top of the page so the user need not scroll back up.

Protocol Links

The **href** attribute of a hyperlink will typically target a resource using the HyperText Transfer protocol (**http:** or secure **https:**) but it may also target resources using other protocols.

Script functions can be called with the **javascript:** protocol, and email clients can be invoked by the **mailto:** protocol:

protocol.html

1 Create an HTML document

2 Within the body section, insert an image of a chart
```
<img id="chart" src="protocol-chart.png" alt="Chart">
```

3 Next, insert a paragraph containing two hyperlinks that target different protocols
```
<p id="links">
<a href="javascript:toggle( )">Show/Hide Chart</a>
<br>
<a href="mailto:wendy@example.com">Email Wendy</a>
</p>
```

4 Now, add a style sheet with a rule to hide the image, and a rule to style the paragraph
```
<style>

img#chart { visibility : hidden ; height : 0px ; }
p#links { padding : 5px ; border : 1px solid ;
                    float : left ; width : 200px ; }

</style>
```

This script first examines the current visibility status of the image element, then reverses it.

5 Finally, add a script to alternately reveal and hide the image whenever the first hyperlink gets clicked
```
<script>

function toggle( ) {
  const chart = document.getElementById( 'chart' )
  let hid = ( chart.style.visibility !== 'visible' )
  chart.style.visibility = ( hid ) ? 'visible' : 'hidden'
  chart.style.height = ( hid ) ? 'auto' : '0px'
}

</script>
```

6 Save the HTML document, then open it in your browser and click on the first link to reveal the chart image

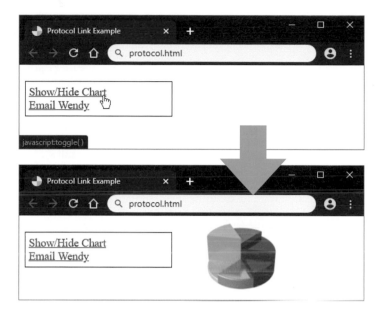

7 Click on the first hyperlink to hide the chart image again

8 Click on the second hyperlink to launch your default client email application – ready to send a message

The **mailto:** protocol automatically adds the email address of the recipient in the "To" field of the email client.

Summary

- Paragraph **\<p\>** elements can include **\<br\>** line break tags, and paragraphs can be separated by **\<hr\>** horizontal ruled lines.

- Long quotations may be enclosed within a **\<blockquote\>** flow element, and short quotations within a **\<q\>** phrasing element.

- The **\<em\>** and **\<strong\>** phrasing elements are preferred over the **\<b\>** and **\<i\>** phrasing elements to emphasize text.

- Side comments can be enclosed within a **\<small\>** element and the **\<ins\>** and **\<del\>** elements used to indicate replaced text.

- The **\<s\>** element denotes superseded content, and the **\<mark\>** element is used to highlight content for reference.

- The **\<u\>** element denotes different text, and the **\<wbr\>** element can be used to suggest an appropriate break point.

- To avoid misalignment, tab spacing should be avoided when creating preformatted text for inclusion within a **\<pre\>** element.

- Superscript and subscript can be included using character entities or using the **\<sup\>** and **\<sub\>** elements.

- Program code can be included in an HTML document using the **\<code\>**, **\<var\>**, and **\<samp\>** elements.

- Machine-readable code can be specified to a **value** attribute of the **\<data\>** tag and to a **datetime** attribute of the **\<time\>** tag.

- The **\<abbr\>**, **\<cite\>**, **\<dfn\>**, and **\<kbd\>** elements provide advice.

- Many elements can include a **title** attribute to provide tooltips.

- The **\<bdi\>** and **\<bdo\>** bi-directionals elements can be used to surround items of text written in a language read right-to-left.

- Ruby annotation uses **\<ruby\>**, **\<rb\>**, **\<rt\>**, **\<rp\>**, and **\<rtc\>** elements to provide pronunciation aid for Eastern languages.

- The **\<a\>** tag can create hyperlinks to other web pages, page fragments, or protocols such as **javascript:** and **mailto:**.

4 Write Lists and Tables

Unordered Lists

Unordered lists, where the sequence of list items is not important, typically place a bullet point before each item to differentiate list items from regular text.

In HTML, unordered lists are created with **<ul> </ul>** tags, which provide a container for list items. Each list item can be created using **<li> </li>** tags to enclose the item, or optionally just using **<li>** to precede the item – either form of **<li>** element validates as correct HTML. An unordered list **<ul>** element can contain numerous list item **<li>** elements.

The bullet point that differentiates unordered list items from regular text may be one of these three marker types:

- **Disc** – a filled circular bullet point (the default style).

- **Circle** – an unfilled circular bullet point.

- **Square** – a filled square bullet point.

A style rule can specify any one of the above values to the unordered list's **list-style-type** property, or a **none** value can be specified to that property to suppress bullet points.

Each HTML list also has a **list-style-image** property that can specify the URL of an image to be used as the list's bullet point. This will appear in place of any of the marker-type bullet points. Where the web browser cannot use the specified image, the marker specified to its **list-style-type** property will be used, or when no marker has been specified, the default will be used.

ulist.html

1 Create an HTML document

2 Within the body section, insert an unordered list that will display the default disc bullet points

```
<ul>

<li>HTML</li>
<li>Cascading Style Sheets</li>
<li>JavaScript</li>

</ul>
```

3 Next, insert an unordered list that will display the circle
bullet points
```
<ul style="list-style-type:circle">
<li>C Programming</li>
<li>C++ Programming</li>
<li>C# Programming</li>
</ul>
```

4 Now, insert an unordered list that will display the square
bullet points
```
<ul style="list-style-type:square" >
<li>Bash<li>PHP<li>Python</ul>
```

5 Finally, insert an unordered list that will display an image
as bullet points
```
<ul style="list-style-image:url(ullst-go.png)" >
<li>Access</li>
<li>Excel VBA</li>
<li>Visual Basic</li>
</ul>
```

ulist-go.png
21px x 21px

6 Save the HTML document, then open it in your browser
to see the unordered list bullet points

Unordered List Example   ×   +

C ⟲ ⌂   🔍 ulist.html

- HTML
- Cascading Style Sheets
- JavaScript

○ C Programming
○ C++ Programming
○ C# Programming

▪ Bash
▪ PHP
▪ Python

go Access
go Excel VBA
go Visual Basic

Note that in CSS terms,
the lists are written in a
content box with their
bullet points drawn in its
left padding area.

Ordered Lists

Ordered lists, where the sequence of list items is important, number each item to differentiate list items from regular text.

In HTML, ordered lists are created with **<ol> </ol>** tags, which provide a container for list items. As with unordered lists, each list item can be created using **<li> </li>** tags to enclose the item, or optionally just using **<li>** to precede the item – either form of **<li>** element validates as correct HTML. An ordered list **<ol>** element can contain numerous list item **<li>** elements.

The automatic numbering that differentiates ordered list items from regular text may be one of these six numbering types:

- **Decimal** – traditional numerals (the default style).

- **Roman** – classical numerals.

- **Latin** – traditional alphabetical lettering.

- **Greek** – classical alphabetical lettering.

- **Georgian** – traditional Georgian numbering.

- **Armenian** – traditional Armenian numbering.

A style rule can specify any of the above numbering types to the list's **list-style-type** property with the following values:

Type	Value
Decimal	**decimal** or **decimal-leading-zero**
Roman	**lower-roman** or **upper-roman**
Latin	**lower-latin** or **upper-latin** **lower-alpha** or **upper-alpha**
Greek	**lower-greek**
Georgian	**georgian**
Armenian	**armenian**

Additionally, a **none** value can be specified to suppress numbering. List item numbering will normally begin at one but a different start point can be specified to a **start** attribute in the **<ol>** tag.

Don't forget

When no numbering type has been specified, the default will be used.

1 Create an HTML document

olist.html

2 Within the body section, insert an ordered list that will display default numbering

```
<ol>

<li>HTML</li>
<li>Cascading Style Sheets</li>
<li>JavaScript</li>

</ol>
```

3 Next, insert an ordered list that will display Roman numbering

```
<ol style="list-style-type:upper-roman">
<li>C Programming</li>
<li>C++ Programming</li>
<li>C# Programming</li>
</ol>
```

4 Now, insert an ordered list that will begin numbering at one hundred (100)

```
<ol start="100" >
<li>Bash<li>PHP<li>Python</ol>
```

5 Save the HTML document, then open it in your browser to see the ordered list numbering

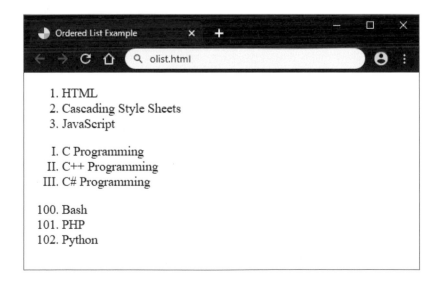

As with the markers in unordered lists, numbering is drawn in the left padding area of the list's content box.

Description Lists

A description list is a unique type of list in which each list item has two parts – the first part being a term, and the second part being a description of the term in the first part. This is referred to as a name/value pair. For example, a name/value pair for the term "sun" could be "sun/the star at the center of our solar system".

In HTML, description lists are created with **<dl> </dl>** tags, which provide a container for list items. Each list item term is contained between **<dt> </dt>** definition term tags, and each list item description is contained between **<dd> </dd>** definition description tags. Optionally, the **</dt>** and **</dd>** closing tags may be omitted – either form of **<dt>** and **<dd>** element is valid.

Each list item in a description list can contain multiple **<dt>** definition term elements and multiple **<dd>** definition description elements – to allow a single term to have multiple descriptions, or multiple terms to have a single description. Typically, browsers display the definition descriptions inset from their terms.

Description lists are also useful to contain a series of questions and related answers, or any other groups of name/value data.

dlist.html

 1 Create an HTML document

2 Within the body section, insert a description list containing two question and answer name/value pairs
```
<dl>
  <dt>What is HTML?</dt>
  <dd>The HyperText Markup Language</dd>

  <dt>What is it used for?</dt>
  <dd>Web page structure.</dd>
</dl>
```

3 Next, in the body section, insert a description list describing the use, pronunciation, and meaning of a term
```
<dl>
<dt><dfn>Homonym</dfn></dt>
<dd class="grammar">noun</dd>
<dd class="spoken">[hom-uh-nim]</dd>
<dd>a word the same as another in sound and spelling
but different in meaning</dd>
</dl>
```

4 Now, insert a description list describing the use, pronunciation, and several meanings of a term

```
<dl>
<dt><dfn>Mouse</dfn></dt>
<dd class="grammar">noun</dd>
<dd class="spoken">[mous]</dd>
<dd>a small animal of various rodent families</dd>
<dd>a palm-sized button-operated device used to move a
computer cursor</dd> <dd>a quiet, timid person</dd>
</dl>
```

The **\<dt\>** element alone does not indicate its content is a term being defined – a nested **\<dfn\>** element must be used for that purpose.

5 Add a style sheet to color the question and definition terms in the lists, and to color some specific descriptions

```
<style>
dt { color : Blue ; }
dfn { color : Red ; font-size : 20pt ; }
dd.grammar { color : Green ; }
dd.spoken { color : Purple ; }
</style>
```

6 Save the HTML document, then open it in your browser to see the name/value pairs

Do not use a definition list to mark up dialog – use paragraphs to mark up each piece of dialog instead.

Description List Example ✕ +

← → C ⌂ Q dlist.html 👤 ⋮

What is HTML?
> The HyperText Markup Language

What is it used for?
> Web page structure.

Homonym
> noun
> [hom-uh-nim]
> a word the same as another in sound and spelling but different in meaning

Mouse
> noun
> [mous]
> a small animal of various rodent families
> a palm-sized button-operated device used to move a computer cursor
> a quiet, timid person

Basic Table

Data is often best presented in tabular form, arranged in rows and columns to logically group related items, so it is easily understood.

In HTML, tables are created with **<table> </table>** tags, which provide a container for table rows. Each table row is created with **<tr> </tr>** tags, which provide a container for a line of table data cells. Each table data cell is created with **<td> </td>** tags, which enclose the actual data to be presented. Optionally, the **</td>** and **</tr>** closing tags may be omitted – either form of **<td>** and **<tr>** element is valid.

A **<table>** element will typically contain numerous **<tr>** elements to create a table displaying multiple rows of data. Similarly, each **<tr>** element will typically contain numerous **<td>** elements to create a table of multiple columns of data. It is important to note, however, that each **<tr>** row in the table must contain the exact same number of **<td>** cells – so, for example, if the first **<tr>** row contains five **<td>** cells, all **<tr>** rows must contain five **<td>** cells.

table.html

1 Create an HTML document

2 Within the body section, insert a table element
```
<table>
<!-- Table rows to go here. -->
</table>
```

Hot tip

Omit the closing **</td>** cell tags but include the closing **</tr>** tags to more clearly denote the end of each table row.

3 Now, within the table element, insert three table rows – that each contain three table data cells
```
<tr> <td>Cell 1.1 <td>Cell 1.2 <td>Cell 1.3 </tr>
<tr> <td>Cell 2.1 <td>Cell 2.2 <td>Cell 2.3 </tr>
<tr> <td>Cell 3.1 <td>Cell 3.2 <td>Cell 3.3 </tr>
```

4 Add a style sheet to set the table width and border, cell borders, and borders of headings that will be added later
```
<style>
table { width : 500px ; border : 5px solid Black ; }
td,th { border : 1px solid Black ; }
</style>
```

5 Save the HTML document, then open it in your browser to see a basic table

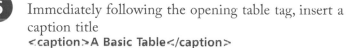

Cell 1.1	Cell 1.2	Cell 1.3
Cell 2.1	Cell 2.2	Cell 2.3
Cell 3.1	Cell 3.2	Cell 3.3

A table title can be specified with **<caption>** **</caption>** tags, and row and column headings can be added between **<th>** **</th>** tags.

 6 Immediately following the opening table tag, insert a caption title
<caption>A Basic Table</caption>

7 Next, insert a new row of four column headings
<tr><th><th>Column 1<th>Column 2<th>Column 3</tr>

8 Now, insert a heading at the start of each following row
<tr><th>Row 1<td>Cell 1.1<td>Cell 1.2<td>Cell 1.3</tr>
<tr><th>Row 2<td>Cell 2.1<td>Cell 2.2<td>Cell 2.3</tr>
<tr><th>Row 3<td>Cell 3.1<td>Cell 3.2<td>Cell 3.3</tr>

9 Save the edited HTML document, then refresh your browser to view the additions

A Basic Table

	Column 1	Column 2	Column 3
Row 1	Cell 1.1	Cell 1.2	Cell 1.3
Row 2	Cell 2.1	Cell 2.2	Cell 2.3
Row 3	Cell 3.1	Cell 3.2	Cell 3.3

Beware

If a **<caption>** element is to be included it must immediately follow the opening **<table>** tag.

Beware

The closing **</th>** tag is optional but the number of opening **<th>** headings must exactly match the number of rows and columns.

Don't forget

Subsequent examples in this chapter build upon this simple table example as more table features are introduced.

Span Cells

An individual table cell can be combined with others vertically to span down over multiple rows of a table.

The number of rows to be spanned is specified to a **rowspan** attribute in the spanning cell's **<td>** tag. Cells in the rows being spanned must then be removed to maintain the table symmetry.

rowspan.html

 Make a copy of the **table.html** document, created in the previous example on pages 84-85, and rename it "rowspan.html"

 Change the document and table titles
<title>Row Spanning Example</title>

<caption>A Table Spanning Rows</caption>

 In the table data element containing the text "Cell 1.1", insert an attribute in its opening tag and edit its content
<td rowspan="2">Cell 1.1+2.1</td>

 Next, delete the table data element containing the text "Cell 2.1" – as this cell is now spanned

5 Now, add rules to the style sheet style to color the background of cells that span rows
td[rowspan="2"] { background : Pink ; }
td[rowspan="3"] { background : HotPink ; }

6 Save the HTML document, then open it in your browser to see the cell spanning two rows in Column 1

	Column 1	Column 2	Column 3
Row 1	Cell 1.1+2.1	Cell 1.2	Cell 1.3
Row 2		Cell 2.2	Cell 2.3
Row 3	Cell 3.1	Cell 3.2	Cell 3.3

A Table Spanning Rows

Row Spanning Example

rowspan.html

7 Next, insert an attribute into the table data element containing the text "Cell 2.2" and edit its content
`<td rowspan="2">Cell 2.2+3.2</td>`

 8 Now, delete the table data element containing the text "Cell 3.2" – as this cell is now spanned

 9 Save the edited HTML document, then refresh your browser to see the cell spanning two rows in Column 2

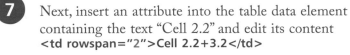

	Column 1	Column 2	Column 3
Row 1	Cell 1.1+2.1	Cell 1.2	Cell 1.3
Row 2		Cell 2.2+3.2	Cell 2.3
Row 3	Cell 3.1		Cell 3.3

A Table Spanning Rows — rowspan.html — Row Spanning Example

10 Next, insert an attribute into the table data element containing the text "Cell 1.3" and edit its content
`<td rowspan="3">Cell 1.3+2.3+3.3</td>`

11 Now, delete the table data elements containing the text "Cell 2.3" and "Cell 3.3" – as these cells are now spanned

12 Save the edited HTML document then refresh your browser to see the cell spanning three rows in Column 3

	Column 1	Column 2	Column 3
Row 1	Cell 1.1+2.1	Cell 1.2	Cell 1.3+2.3+3.3
Row 2		Cell 2.2+3.2	
Row 3	Cell 3.1		

A Table Spanning Rows — rowspan.html — Row Spanning Example

Don't forget

Notice that by default text in each cell is left-aligned and horizontally centered in merged cells.

Enhance Tables

Tables can be enhanced by the addition of special header and footer rows above and below the regular table content, which provide additional table information.

In HTML, table header information is contained between **<thead>** **</thead>** tags, and table footer information is contained between **<tfoot>** **</tfoot>** tags. When a table has a **<thead>** and/or a **<tfoot>** element, all regular table rows must be enclosed between **<tbody>** **</tbody>** tags.

In long tables, rows can be grouped into separate table body sections using multiple **<tbody>** elements. When these are printed, each paper page can repeat the table header and footer information.

It is important to note that a **<thead>** element must appear before the first **<tbody>** element within the **<table>** element, but after the **<caption>** element if one is present.

enhance.html

 Make a copy of the **table.html** document, created on pages 84-85, and rename it "enhance.html"

 Change the document and table titles
<title>Enhanced Table Example</title>

<caption>An Enhanced Table</caption>

3 Add rules to the style sheet to style a table header, a second table body heading, and a final table footer
thead { background : Pink ; }
th.next { background : DeepPink ; color : White ; }
tfoot { background : HotPink ; }

4 Immediately after the caption, insert a table header containing a single row that spans all four columns
<thead>
<tr><td colspan="4">Header Information</tr>
</thead>

5 After the header, add a table body element to enclose all the regular existing table rows
<tbody>
<!-- Existing row elements go here. -->
</tbody>

6 After the table body element, insert a second table body element containing four more table rows

```
<tbody>

<tr><th colspan="4" class="next">Next section</tr>
<tr>
<th>Row 4<td>Cell 4.1<td>Cell 4.2<td>Cell 4.3</tr>
<tr>
<th>Row 5<td>Cell 5.1<td>Cell 5.2<td>Cell 5.3</tr>
<tr>
<th>Row 6<td>Cell 6.1<td>Cell 6.2<td>Cell 6.3</tr>

</tbody>
```

7 After the second table body, insert a table footer containing a single row that spans all four columns

```
<tfoot>
<tr><td colspan="4">Footer Information</tr>
</tfoot>
```

8 Save the HTML document, then open it in your browser to see the enhanced table

An Enhanced Table			
Header Information			
	Column 1	**Column 2**	**Column 3**
Row 1	Cell 1.1	Cell 1.2	Cell 1.3
Row 2	Cell 2.1	Cell 2.2	Cell 2.3
Row 3	Cell 3.1	Cell 3.2	Cell 3.3
		Next section	
Row 4	Cell 4.1	Cell 4.2	Cell 4.3
Row 5	Cell 5.1	Cell 5.2	Cell 5.3
Row 6	Cell 6.1	Cell 6.2	Cell 6.3
Footer Information			

Don't forget

Table headers and footers should only contain information   all table data should appear in the table body.

Control Columns

Where a table simply has an overall width specified by a style rule, the browser will by default calculate the width of each column according to its content – columns with broad content will be wider than columns with slender content. Greater control over column width can be achieved using **<col>** tags to represent individual columns, so rules can specify their size and appearance.

A single **<col>** element can also represent multiple columns by including a **span** attribute to specify a number of columns. So a style rule specifying a column width will be applied to all the columns that **<col>** element represents. Alternatively, the **<th>** or **<td>** tags can include a **colspan** attribute to specify a number of columns to span.

Optionally, **<col>** elements may be enclosed between **<colgroup> </colgroup>** tags to allow styling of both column groups and individual columns.

column.html

 Create an HTML document

 Within the body section, insert a table element that includes a caption
<table>
<caption>Breakfast Flights</caption>

<!-- Table content to go here. -->

</table>

 Next, in the table, insert a column group that includes a class name for styling and contains a single column
<colgroup class="sidebar">
<col>
</colgroup>

 Now, insert two more column groups that include class names for both group styling and individual styling
<colgroup class="info">
<col class="stripe"> <col> <col class="stripe">
</colgroup>

<colgroup class="info">
<col> <col class="stripe">
</colgroup>

5 After the column groups, insert a table header, a table body, and a table footer – each with six columns

```
<thead><tr><th colspan="6"><!-- Header. --></thead>

<tbody><!-- Rows with six cells each. --></tbody>

<tfoot><tr><td colspan="6"><!-- Footer. --></tfoot>
```

6 Add a style sheet with rules to specify the appearance of the table, and its header, footer, and data cells

```
</style>
table { width : 500px ; border-collapse : collapse ; }
tbody th { background : DeepPink ; color : White ; }
tbody td { padding : 3px ; text-align : center ; }
tfoot { font-size : small ; }
</style>
```

7 Next, add rules to specify the width of each column

```
colgroup.sidebar col { width : 70px ; }
colgroup.info col { width : 80px ; }
```

8 Now, add rules to style groups and individual columns

```
colgroup.info { border-left : 2px solid White ; }
colgroup col.stripe { background : Pink ; }
```

9 Save the HTML document, then open it in your browser to see distinct column groups

---

Column Styling Example   ✕   +

← → C ⌂   Q column.html   ⊖ ⋮

Breakfast Flights
New York (JFK) - Los Angeles (LAX)

	American Airlines	Delta Air Lines	Alaska Airlines	United	Continental
Departure	08:30	07:00	07:30	07:55	08:35
Arrival	12:05	10:30	10:45	11:30	12:00
Duration	6h35min	6h30min	6h15min	6h35min	6h25min
Price	$179	$195	$235	$225	$189

• Flights are Non-Stop • Times are Local • Tickets are 1-Way • Prices Include Tax

Hot tip

The **•** character entity is used in this table footer to create bullet points.

Summary

- The HTML **<ul>** element creates an unordered bullet point list that contains individual list items within **<li>** elements.

- A **list-style-type** property can specify that unordered list items should have a **disc**, **circle**, or **square** bullet point, or **none**.

- A **list-style-image** property can specify the URL of an image that should appear in place of list item bullet-points.

- The **<ol>** element creates an ordered numerical list that contains individual list items within **<li>** elements.

- A **list-style-type** property can specify how ordered list items should be numbered, such as **decimal**, **upper-latin**, or **none**.

- The **<dl>** element creates a definition list containing terms in **<dt>** elements, and their descriptions in **<dd>** elements.

- The HTML **<table>** element creates a table, and may optionally first enclose a **<caption>** element to title the table.

- Each table row is created with a **<tr>** element to contain numerous **<th>** heading elements and **<td>** data elements.

- Table cells can span down other cells using the **rowspan** attribute, and cells to the right using the **colspan** attribute.

- Adding **<thead>** and **<tfoot>** elements immediately after the **<caption>** element enhances a table with a header and footer.

- Tables that have a header and footer must also enclose all regular table rows within a **<tbody>** element.

- Table columns can be grouped using a **<colgroup>** element for styling.

- Each table column can be represented by a **<col>** element so it can be individually styled.

5 Incorporate Media Content

This chapter demonstrates how to include images and other media in page content.

Add Images

The ability to add images to HTML document content introduces lots of exciting possibilities. An image is easily added to the document using the **<img>** tag, which should preferably always include these attributes:

- A **src** attribute is required to specify the image location URL, by either its absolute or relative path.

- A **width** attribute is recommended to specify the pixel width of the area the image will occupy on the page.

- A **height** attribute is recommended to specify the pixel height of the area the image will occupy on the page.

- An **alt** attribute is recommended to specify text describing the image, for occasions when the image cannot be loaded.

Attributes in HTML tags can appear in any order.

The values assigned to the **width** and **height** attributes instruct the web browser to create a content area on the web page of that size. This need not be the actual dimensions of the image, as the web browser can render the image in another specified size. Care must be taken to avoid distortion by ensuring the dimensions are scaled in proportion to the actual image size. Additionally, images should only be scaled down, as scaling up often results in pixelation – where individual pixels are visible to the eye. It is inefficient, however, to rely upon the browser to scale images that are not to be displayed full size, as this requires downloading unnecessarily larger files. It is better to adjust the image size to the actual dimensions it will occupy on the web page using a graphics editor, such as Adobe Photoshop, so it will download and display faster.

Avoid the BMP bitmap file format for web graphics – saving the original image shown here as **fish.bmp** creates a file size of 790KB!

Original file size

Item type: PNG File
Dimensions: 600 x 450
Size: 189 KB

Reduced to 33%

Item type: PNG File
Dimensions: 200 x 150
Size: 30.5 KB

The optimum file type for web bitmap graphics is the popular non-proprietary Portable Network Graphics (PNG) format, which produces compact files and supports transparency.

1 Create an HTML document

image.html

2 Within the body section, insert three image elements – to display a graphic at full size plus two scaled versions
<img src="image-fish.png"
  width="200" height="150" alt="Fish">
<img src="image-fish.png"
  width="150" height="112" alt="Fish">
<img src="image-fish.png"
  width="100" height="75" alt="Fish">

3 Save the HTML document then open it in your browser to see the full-size image and the two scaled versions

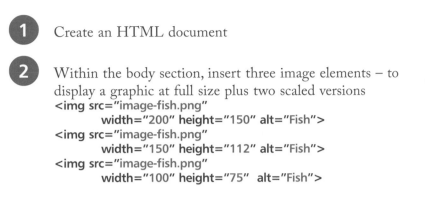

image-fish.png
200px x 150px

4 Now, insert this attribute into each image element
style="background:Aqua"

5 Refresh your browser to see the colored backgrounds

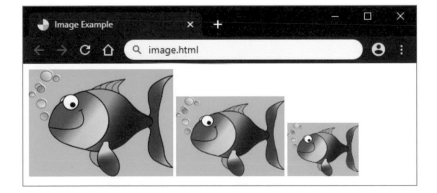

Image Maps

A single image can target multiple hyperlink resources if an image "map" is added to define "hot spot" areas for each hyperlink. To use an image map, the **<img>** tag must include a **usemap** attribute to specify a map name, prefixed by a **#** hash character. The image map itself is contained between **<map> </map>** tags, and its name is specified by a **name** attribute in the opening **<map>** tag.

Each area of the image that is to become a hyperlink hot spot is defined by four attributes of an **<area>** tag within the **<map>** element. The **shape** attribute specifies its shape as **rect** (rectangle), **circle**, or **poly** (polygon), and the **coords** attribute specifies a comma-separated list of its x-axis and y-axis coordinates:

Shape	Coordinates
rect	top-left x, top-left y, bottom-right x, bottom-right y
circle	center x, center y, radius
poly	x1, y1, x2, y2, x3, y3, etc. – one pair for each point. The first and final point must have identical coordinates to complete the shape

Additionally, each **<area>** tag must have an **href** attribute, to specify the hyperlink's URL target, and an **alt** attribute to specify alternative text to be displayed when images are not enabled.

1 Create an HTML document

2 Within the body section, insert an image and map element
```
<img src="map.png" alt="Search" usemap="#search">
<map name="search">

   <!-- Areas to go here. -->

</map>
```

map.html

map.png
400px x 200px

3 Within the map element, define a rectangular hot spot covering the top-left quarter of the image – from a top-left point at xy:0,0 to a bottom-right point at xy:200,100
```
<area   shape="rect" coords="0,0,200,100"
        href="https://www.bing.com"
        alt="Bing Panel" title="Link to Bing">
```

4 Now, in the map element, define three hot spots of the same size covering the other three quarters of the image

```
<area   shape="rect" coords="200,0,400,100"
        href="https://www.ask.com"
        alt="Ask Panel" title="Link to Ask">

<area   shape="rect" coords="0,100,200,200"
        href="https://www.google.com"
        alt="Google Panel" title="Link to Google">

<area   shape="rect" coords="200,100,400,200"
        href="https://www.yahoo.com" alt="Yahoo Panel"
        title="Link to Yahoo">
```

5 Save the HTML document, then open it in your browser to see the tooltips describe each hot spot that you can click to open its associated target

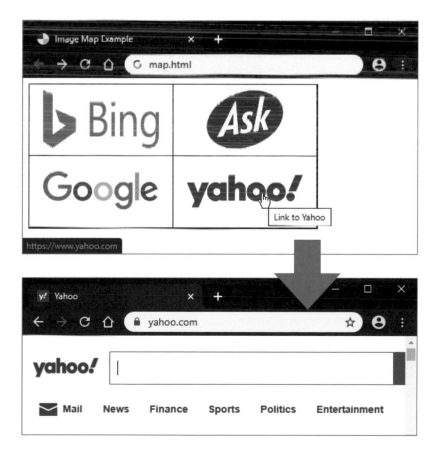

Do not leave any spaces in the comma-separated list of coordinates.

Validation will fail unless each **<area>** tag includes an **alt** attribute.

Reference Figures

With the latest HTML specifications, the web page author now has additional means by which to insert images into a web page. An **<img>** tag can be nested within a **<figure> </figure>** element to embed an image that is related to the main text content but whose removal would not disrupt the text's meaning.

As the nested image, in effect, is now self-contained as a "figure" it can be positioned away from the text if desired, and referenced by a caption within a nested **<figcaption> </figcaption>** element:

figure.html

 Create an HTML document

2 Within the body section, insert a heading
<h1>Web Development Stacks</h1>

3 Next, insert a captioned figure
<figure id="front-stack" >

<img src="figure-front.png" alt="Front-end"
width="160" height="145">
<figcaption class="reference">
Figure 1:Front-end Technologies
</figcaption>

</figure>

4 Now, insert text content that makes reference to the previous captioned figure
<p>Front-end development, also known as client-side development, is the practice of producing HTML documents, CSS style sheets, and JavaScript script code <span class="reference" >(Figure 1)</span> for a website or Web Application - so a user can see and interact with them directly.</p>

5 Insert a second captioned figure
<figure id="back-stack" >

<img src="figure-back.png" alt="Back-end"
width="160" height="128" >
<figcaption class="reference" >
Figure 2:Back-end Technologies
</figcaption>

</figure>

Always refer to figures only by their label – avoid using reference terms like "in the figure on the right" so the document layout can be easily changed without creating confusion.

6 Now, insert text content that makes reference to the
second captioned figure
```
<p>Back-end development, also known as server-side
development, is the practice of producing complex
websites using programming languages such as SQL, Java,
PHP, or .NET <span class="reference" >(Figure 2)</span>
to provide features beyond front-end capabilities.</p>
```

7 Add a style sheet to position each captioned figure and to
specify some font styles
```
<style>
figure#front-stack { float : left ; margin-top : 0px ; }
figure#back-stack { float : right ; margin-top : 0px ; }
.reference { color : Red ; font-weight: bold ; }
p:first-letter { font-size : xx-large ; }
</style>
```

8 Save the HTML document and style sheet then open the
web page in your browser to see the captioned figures

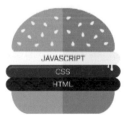

Web Development Stacks

JAVASCRIPT
CSS
HTML

Figure 1:Front-end Technologies

Front-end development, also known as client-side development, is the practice of producing HTML documents, CSS style sheets, and JavaScript script code (Figure 1) for a website or Web Application - so a user can see and interact with them directly.

Back-end development, also known as server-side development, is the practice of producing complex websites using programming languages such as SQL, Java, PHP or .NET (Figure 2) to provide features beyond front-end capabilities.

SQL
JAVA/PHP, .NET

Figure 2:Back-end Technologies

Don't forget

You can discover more about back-end technologies with the companion books in this series on SQL, Java, PHP, Python, and MySQL at www.ineasysteps.com

Select Pictures

As web content is increasingly being accessed on small handheld devices, the latest HTML specifications allow the web page author to specify alternative images to be displayed on the web page according to the size of the device screen.

A **<picture>** **</picture>** element is used to contain multiple image sources from which the browser can select the most appropriate size. Each image source is specified to the **srcset** attribute of a nested **<source>** element, and the minimum screen width suitable for that image is specified to its **media** attribute. The assignment requires an unusual syntax that states the size to a **min-width** property within **( )** parentheses – for example, to specify that an image is suitable for display only on devices whose screen width exceeds 500 pixels with **media="(min-width : 500px)"**.

Usefully, the **<picture>** **</picture>** element can enclose a final regular **<img>** element to specify the image to be displayed on older web browsers that do not support this selection feature:

picture.html

 Create an HTML document

2 Within the body section, insert a container element **<picture>**

<!-- Image sources to be inserted here -->

</picture>

3 Next, insert an image source for display only on devices whose screen width exceeds 500 pixels
<source media="(min-width : 500px)"
       srcset="picture-large.png" >

4 Now, insert an image source for display only on smaller devices whose screen width exceeds 200 pixels
<source media="(min-width : 200px)"
       srcset="picture-small.png" >

 Finally, insert an image source for display only on older browsers that do not support the selection feature
<img   alt="Regular Guy"
       src="picture-medium.png"
       width="250" height="250">

The **<source>** element does not require a closing tag. It is also used within the **<audio>** element (see page 108), and with the **<video>** element (see page 110).

6 Save the HTML document, then open it in various browsers to see an appropriately-sized image

Embed Objects

An external resource can be embedded into an HTML document using **<object>** **</object>** tags to define the resource. When the resource is an image it will be treated much like those specified by **<img>** elements, otherwise a plugin may be sought to process the resource. The **<object>** element can specify the resource's URL to its **data** attribute, and the resource type to its **type** attribute. The resource type must be a valid MIME type describing the resource.

This table lists some popular MIME types. Further details can be found on the W3C website at **www.w3.org**

MIME Type	Object File Format
image/png	PNG image resource
image/jpeg	JPG, JPEG, JPE image resource
image/gif	GIF image resource
image/svg+xml	SVG vector image resource
text/plain	TXT regular plain text resource
text/html	HTM, HTML markup text resource
application/pdf	PDF portable document resource
application/msword	DOC Word document resource
application/x-java-applet	CLASS Java applet resource
audio/x-wav	WAV sound resource
audio/mpeg	MP3 music resource
video/mp4	MP4 video resource
video/x-mpeg	MPEG, MPG, MPE video resource
video/x-msvideo	AVI video resource
video/x-msv-wmv	WMV Windows video resource
video/quicktime	MOV Quicktime video resource

All **<object>** elements must contain at least one **data** attribute or one **type** attribute.

Each **<object>** element can specify dimensions in which to display visual content using its **width** and **height** attributes. Where the resource is an image, the **<object>** element can also include a **usemap** attribute to specify the name of an image map, just like those produced for an **<img>** element.

Optionally, fallback text can be included between the **<object>** **</object>** tags that will only be displayed by the browser in the event that the resource cannot be embedded within the document – for example, when an appropriate plugin cannot be found.

...cont'd

1 Create an HTML document

2 Within the body section, insert a paragraph wrapper
**<p>This is text in the main document that...
**

<!-- Resource object to be embedded here. -->

**
...continues around an embedded resource.</p>**

object.html

3 Within the paragraph, insert a PDF object to embed
<object data="object-chart.pdf" type="application/pdf"
            width="500" height="350">
[PDF Document - May require the Adobe Reader plugin]
</object>

object-chart.pdf
(external resource)

4 Save the HTML document alongside the specified resource file then open the web page in your browser to see the embedded object

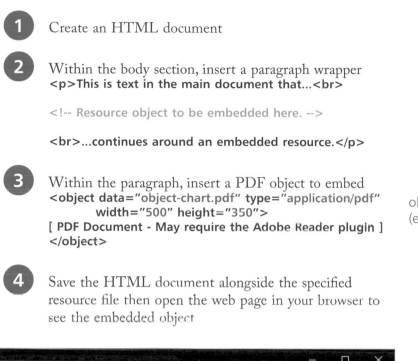

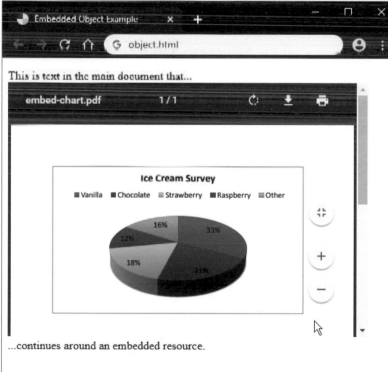

Embed Vectors

The HTML **<embed>** element allows you to integrate an external resource for interaction with the HTML document.

Web browsers that support modern HTML also support Scalable Vector Graphics (SVG). Unlike bitmap graphic formats such as PNG, which store their graphic information as the color of each pixel, vector graphics store the graphic information as a series of "paths". This is a highly efficient way to describe graphics.

Most importantly, vector graphics can be scaled without loss of fidelity. This means that they can be infinitely enlarged without suffering the pixelation experienced when enlarging bitmap images, such as those in PNG, JPG, or GIF file formats.

SVG Vector x 3   PNG Bitmap x 3

SVG is not actually part of HTML but is a specification based on the eXtensible Markup Language (XML), so it describes vector images in text files. These can be created manually but it's far simpler to use a vector graphics editor such as Adobe Illustrator.

SVG images can be zoomed without loss of definition and can be printed in high quality without loss of resolution.

SVG files can also be scripted. Adding JavaScript functionality to a static vector image makes it possible to create interactive SVG objects. This means that every element and every attribute within an SVG file can be animated.

Both static SVG images and interactive SVG objects can be embedded in HTML by specifying the MIME type of "image/svg+xml" to the **<embed>** element's **type** attribute.

Hot tip

Static SVG images can alternatively be embedded using the **<img>** element – just like any other image.

1 Create an HTML document

2 In the body section, insert an element to embed an interactive scalable vector graphic
```
<embed src="vector-picker.svg" type="image/svg+xml"
         width="280" height="200" >
```

3 Save the HTML document alongside the SVG file, then open the web page in your browser to see the embedded interactive scalable vector graphic

4 Pick a color by clicking any circled color sample in the image to interact with the vector graphic

vector.html

vector-picker.svg
(external resource)

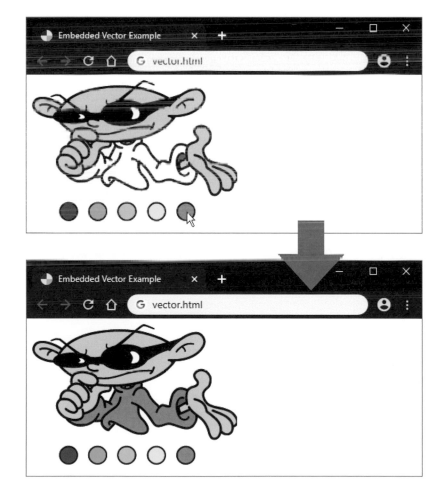

You can examine how JavaScript has been incorporated into this SVG document by downloading the examples archive from: **www.ineasysteps. com/resource-centre/ downloads**

Embed Frames

External resources can be embedded in an HTML document within an "inline frame" using **<iframe> </iframe>** tags. These create a fixed area on the page in which to display the embedded resource. The inline frame's dimensions must be specified to the **<iframe>** element's **width** and **height** attributes, and the URL of the external resource to its **src** attribute. Where the dimensions of the external resource exceed those of the inline frame, the browser automatically adds scroll bars so the user can view the entire content.

Each **<iframe>** element may also optionally contain a **name** attribute to specify a unique identifier for that frame. This allows hyperlinks to then load the URL specified to their **href** attribute into the inline frame (rather than replace the entire page) by assigning the frame name to a **target** attribute in the **<a>** element. For example, a hyperlink could target an inline frame named "topbox" with **<a href="some.html" target="topbox">**.

Typically, inline frames are useful to provide supplemental content while maintaining a compact page format.

iframe.html

 Create an HTML document

 Within the body section, insert an article containing a heading and descriptive paragraph, and with a specified class name for positional styling purposes
<article class="left220">

<h3>Concept Cars</h3>
<p>Many of the creative and innovative concept cars premiered at the recent motor show left the audience in eager anticipation of their production.</p>

</article>

concept.html
(external resource)

3 Next, in the body, as an aside, insert an inline frame to load a document containing relevant text and illustrative photographs positioned horizontally side-by-side
<aside>

<iframe src="concept.html" width="300" height="220">

</iframe>

</aside>

...cont'd

4 Now, add a style sheet to size the article and position it to the left of the inline frame

```
<style>
article.left220 { width : 200px ; float : left ;
                  margin-right : 10px ; }
</style>
```

5 Save the HTML document alongside the external resource, then open it in your browser to see the article and the inline frame content

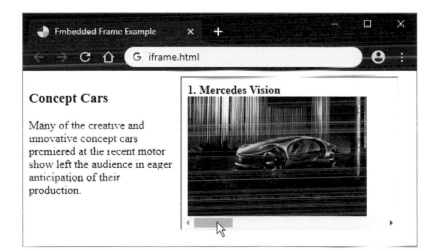

A fallback message can be provided between the <iframe> </iframe> tags to be displayed when inline frame support is disabled.

6 Drag the scrollbar to advance through the content

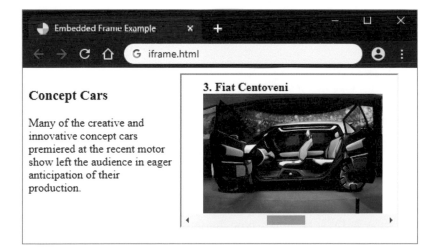

Embedding documents within inline frames is particularly favored on property websites to accompany property descriptions with photographs in a compact page format.

Add Audio

External audio resources such as MP3 music files can be embedded in an HTML document using **<audio> </audio>** tags.

The **<audio>** element can include a **src** attribute to specify the URL of the audio resource to embed, and may include additional attributes to determine how the audio resource will be used:

- **autoplay** – a boolean attribute that specifies the browser should immediately begin playing the audio resource.

- **loop** – a boolean attribute that specifies the browser should play the audio resource repeatedly.

- **controls** – a boolean attribute that specifies the browser should display user controls to start or stop the audio playing.

- **preload** – accepts values of "auto" or "none" to suggest the browser should load the audio resource so it is ready to play.

Boolean attributes, like the **autoplay, loop,** and **controls** attributes, need have no assigned value – their presence alone within the element is sufficient for the browser to understand their purpose.

Browsers rely upon an in-built "codec" (**co**der-**dec**oder) to decode audio resources so they can be played. Sadly, not all browsers incorporate the same audio codec:

- **Advanced Audio Coding (AAC)** – codec "mp4a.40.2" supported by modern browsers such as Google Chrome, Firefox, and Microsoft Edge for MP3 audio.

- **Ogg audio** – codec "vorbis" supported by other browsers for audio files in OGG format.

This inconsistency therefore requires audio resources to be encoded twice for playback across all browsers. Two **<source>** elements may be nested within an **<audio>** element for this purpose, rather than specifying a single resource URL to a **src** attribute in the **<audio>** tag. For each file format, the **<source>** elements can then specify their resource URL to a **src** attribute, and their MIME type to a **type** attribute. The browser will only load the supported audio resource for playback.

A boolean value can be only True or False. By default, attributes that represent boolean values are True unless they are assigned a value of False.

A fallback message can be included between the **<audio> </audio>** tags to be displayed when audio playback support is disabled.

...cont'd

1 Create an HTML document

2 In the body section, insert an element to embed an audio resource in the MP3 format for automatic playback
```
<audio src="audio.mp3" autoplay >
[ Fallback Message ]
</audio>
```

audio.html

3 Save the HTML document then open the web page to hear automatic audio playback in supported browsers

4 Next, replace both previous attributes with one to display user controls for audio playback
```
<audio controls>

<!-- Sources to be inserted here. -->

</audio>
```

Avoid automatic audio playback on websites as many users detest the autoplay feature.

5 Now, in the audio element, insert elements to specify audio resources to be embedded for all browsers
```
<source src="audio.mp3" type="audio/mpeg" >
<source src="audio.ogg" type="audio/ogg" >
```

audio.mp3   audio.ogg
(external resources)

6 Save the HTML document again, then open the web page in any browser and use the controls to hear playback

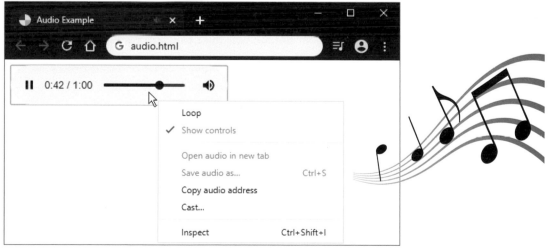

Add Video

External video resources such as MP4 video files can be embedded in an HTML document using **<video>** **</video>** tags.

To determine how the video resource will be used, the **<video>** element can include **src**, **autoplay**, **loop**, **controls**, and **preload** attributes, just like the **<audio>** element in the previous example on pages 108-109. Additionally, the dimensions of the area in which to display the video on the page can be specified to **width** and **height** attributes.

There are two main video compression standards:

- **Advanced Video Coding (AVC)** – a patented standard that is also known as H.264 or MPEG-4 (**.mp4** files).

- **WEBM video** – a royalty-free alternative to the patented H.264 and MPEG-4 standard (**.webm** files).

Video resources can be encoded in each format for playback across all browsers and embedded using **<source>** elements nested within a **<video>** element. For each file format, the **<source>** elements can then specify their resource URL to a **src** attribute, and the MIME type of each video file can be specified to the **type** attribute. The browser will only load the supported video resource for playback.

A **<track>** tag may be nested within a **<video>** element to specify the location of a Web Video Text Tracks (WebVTT) subtitles file to its **src** attribute. This may include a **kind** attribute to describe the track and **srclang** to describe the language. The tag must also include a boolean **default** attribute to use the specified file.

The subtitles file begins with **WEBVTT**. Start and end timing cues are added on new lines in the format HH:MM:SS.sss and separated by **-->** . The associated subtitle caption appears on a new line below each timing cue, like this:

WEBVTT

00:00:01.000 --> 00:00:04.000
Playing Guitar with "HTML in easy steps"

00:00:05.000 --> 00:00:06.000
Thanks for watching

You can discover more about the WebVTT subtitle format online at www.w3.org/TR/webvtt1

Note that the milliseconds are separated by a period (full stop) – not a colon.

video.vtt

Create an HTML document

video.html

 In the body section, insert an element to display user controls for video playback
<video controls >
<!-- Sources to be inserted here. -->
[Fallback Message]
</video>

 Next, in the video element, insert elements to embed a video resource and to specify a subtitle file
<source src="video.mp4" type="video/mp4" >
<source src="video.webm" type="video/webm" >
<track src="video.vtt"
　　　　kind="subtitles" srclang="en" default>

video.mp4　　video.webm
(external resources)

 Save the HTML document then open the web page in any browser and use the controls to see video playback

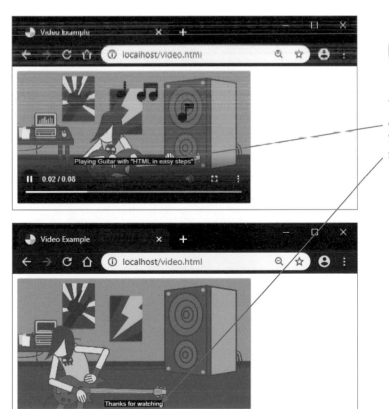

Don't forget

This short video displays a subtitle for 4 seconds then later displays a second subtitle.

Indicate Progress

If you prefer not to provide the browser's standard controls for playback of audio or video, the **controls** attribute can be omitted from the **<audio>** and **<video>** tags. The JavaScript **play( )** and **pause( )** methods of an embedded media object can then be called to control playback from an **onclick** event-handler script function.

A visual indicator of media playback can be displayed using a **<progress> </progress>** element to present a "progress bar". Within the **<progress>** tag, a **value** attribute determines the extent of progress towards completion. This can be dynamically updated in synchronization with media playback from an **ontimeupdate** event-handler script function.

Embedded media objects have a **currentTime** property, which stores the elapsed time since playback began, and a **duration** property that stores total playback time. These can be used to calculate playback progress as a percentage:

progress.html

audio.mp3   audio.ogg
(external resources)

Hot tip

Include an **id** attribute in the **<audio>** tag to reference the media from script, and in all other tags the script needs to reference.

1 Create an HTML document

2 In the body section, insert elements to embed an audio resource for manual playback
```
<audio id="snd">

<source src="audio.mp3" type="audio/mpeg" >
<source src="audio.ogg" type="audio/ogg" >
[ Fallback Message ]

</audio>
```

3 Next, insert an image button to control payback
```
<img id="ctl" src="progress-button.png"
        width="32" height="32" alt="Control" >
```

4 Now, insert elements to present a visual indicator and calculated percentage as playback proceeds
```
<progress id="bar" value="0"></progress>

<span id="num">[Audio]</span>
```

5 Add a script with a function to initialize variables when the HTML document has loaded

```
<script>
( function ( ) {

  const snd = document.getElementById( 'snd' )
  const ctl = document.getElementById( 'ctl' )
  const bar = document.getElementById( 'bar' )
  const num = document.getElmentById( 'num' )
  let run = true

  /*  Event-handler functions go here. */

} ) ( )
</script>
```

The control will play or pause playback according to the boolean value of the **run** variable.

6 Insert a function to control playback

```
ctl.onclick = function( ) {
  ( run ) ? snd.play( ) : snd.pause( )
  run = !run }
```

The **currentTime** and **duration** properties store time in seconds as floating-point values, so need to be rounded down with **Math.floor( )**.

7 Insert a function to display playback progress

```
snd.ontimeupdate = function( ) {
  bar.value = ( snd.currentTime / snd.duration )
  num.innerHTML = Math.floor( 100 * bar.value) + '%' }
```

8 Save the HTML document, then open it in a browser and click the button to see playback progress

Use Templates

The HTML **<template> </template>** element allows you to include content that is not be displayed on the page immediately when the web page loads into the browser.

Content that is stored inside a **<template>** element can be loaded later using JavaScript to display the content on the web page. This is useful when you have content that may be added repeatedly.

Template content is a "DocumentFragment" object that the script can copy. The copy can then be appended as a "child" element of existing content. A script can also remove child elements from content.

template.html

star.png32px x 32px
(gray area is transparent).

 Create an HTML document

 Within the body section, add two paragraphs that will call script functions when clicked
```
<p id="hotel" style="cursor:pointer"
        onclick="addStar( this )">Hotel Mistrale</p>

<p id="clear" style="cursor:pointer"
        onclick="removeStars( )">Clear</p>
```

 Next, add template content that will be appended to existing content by a script function
```
<template>

<img src="star.png" alt="Gold Star">

</template>
```

4 Now, add a script function to copy the template content and append it to the first paragraph when clicked
```
<script>

function addStar( hotel ) {
  const temp = document.getElementById( 'star' )
  const copy = temp.content.cloneNode( true )
  hotel.appendChild( copy )
}

// Second function to be inserted here.

</script>
```

5 Then, insert a function to remove all child elements from the first paragraph when the second paragraph is clicked

```
function removeStars( ) {
  const hotel = document.getElementById( 'hotel' )
  while ( hotel.ChildElementCount > 0 )
  {
    hotel.removeChild( hotel.lastChild )
  }
}
```

6 Save the HTML document alongside the star image, then open it in your browser and click to add/remove content

Hot tip

Before the **<template>** tag was introduced into HTML, the script would have to create the HTML child element content, but using a template is much more convenient.

Insert Slots

The **<template>** element, demonstrated in the previous example on pages 114-115, can only append the content defined within the template element. A template can be made more flexible, however, by including HTML **<slot> </slot>** elements within the **<template>** element, whose content can differ in each instance of the template.

The **<slot> </slot>** tags can, optionally, enclose default text, but the opening **<slot>** tag must include a **name** attribute that will identify that slot as a placeholder within the template.

The **<slot>** element placeholders are filled with content by including a **slot** attribute within an HTML element that nominates the matching **name** attribute value of the slot. JavaScript can then be used to append the combined template and slot content to the web page.

slot.html

 1 Create an HTML document

2 Within the body section, add two divisions that each enclose three spans nominating the same three slots
```
<div class="homonym">
<span slot="word">Air</span>
<span slot="def-1">A lilting tune</span>
<span slot="def-2">What we breathe</span>
</div>

<div class="homonym">
<span slot="word">Current</span>
<span slot="def-1">A flow of water</span>
<span slot="def-2">Up to date</span>
</div>
```

3 Next, add a template of a description list whose terms and descriptions provide the nominated slots
```
<template id="list-template">
<dl>
<dt><slot name="word">Term</slot>
<dd><slot name="def-1">1st Definition</slot>
<dd><slot name="def-2">2nd Definition</slot>
</dl>
<style>
dl { width : 250px ; border : 1px solid ; }
dt { background : Orange ; color : White ; }
</style>
</template>
```

Beware

The style sheet must be included in the template or its rules will not be applied to the template's elements.

...cont'd

4 Now, add a script to combine the list template with the text content of the nominated slots when the page loads
<script>

```
( function ( ) {
  const homs =
          document.getElementsByClassName( 'homonym' )
  const temp = document.getElementById( 'list-template' )

  if( 'attachShadow' in homs[ 0 ] )
  {
    let i, copy, shadow

    for( i = 0 ; i < homs.length ; i++ )
    {
      copy = temp.content.cloneNode( true )
      shadow = homs[ i ].attachShadow( { mode: 'closed' } )
      shadow.appendChild( copy )
    }

  }

} ) ( )
```

</script>

5 Save the HTML document, then open it in your browser to see the combined template and slot content

```
┌─ Slot Example ──── x   + ──────── □ ×
← → C ⌂  G  slot.html            ⊖ ⋮
┌──────────────────────────────┐
│Air                            │
│    A lilting tune             │
│    What we breathe            │
│                               │
│Current                        │
│    A flow of water            │
│    Up to date                 │
└──────────────────────────────┘
```

Hot tip

Strictly speaking, this example creates a "Shadow Document Object Model" (ShadowDOM) that the script then appends to the original document. The **closed** mode simply prevents scripting access to the ShadowDOM via the **shadowRoot** property of the HTML element.

Don't forget

The **slot** attribute can only appear in these HTML tags:
<article>
<aside>
<blockquote>
<body>
<div>
<footer>
<h1-6>
<header>
<main>
<nav>
<p>
<section>
<span>

Employ Dialogs

You can create a "modal" dialog, to which the user must respond before regaining access to the web page, by enclosing the dialog's content between **<dialog>** **</dialog>** tags.

A modal dialog appears on a layer above all other page content and can contain all types of content (text, images, etc.), but must provide some means of closing the dialog to return to the page. Typically, this will be provided by including one or more buttons on the modal dialog that will execute a script function when clicked to perform some action and to close the dialog.

A clickable button can be added to a web page or modal dialog with a **<button>** element. Text between **<button>** and **</button>** tags will appear on the button as its label. The opening **<button>** tag can include an **onclick** attribute to nominate a script function to be called when the button is clicked. Alternatively, the script can dynamically add an "event listener" for each button to recognize when the user clicks a button – creating a "click event".

dialog.html

 Create an HTML document

Within the body section, add a button and an empty paragraph that each have a unique **id** for scripting
<button id="show">Show Dialog</button>
<p id="info"></p>

Next, add a dialog containing text, an image, and two buttons that each have a unique **id** for scripting
<dialog id="dlog">

Your Choices

dialog-qmark.svg

```
<img src="dialog-qmark.svg"
        height="64" width="64" alt="Question Mark">
<br>
<button id="cncl">Cancel</button>
<button id="conf">Confirm</button>

</dialog>
```

4 Now, add a script to add event listeners and event handler functions for each button

```
<script>

( function ( ) {
  const dlog = document.getElementById( 'dlog' )
  const info = document.getElementById( 'info' )
  const show = document.getElementById( 'show' )
  const cncl = document.getElementById( 'cncl' )
  const conf = document.getElementById( 'conf' )

  show.addEventListener( 'click',  function ( ) {
    dlog.showModal( )
    info.innerText = 'Modal Dialog Open' } )

  cncl.addEventListener( 'click',  function ( ) {
    dlog.close( )
    info.innerText = 'Modal Dialog Canceled' } )

  conf.addEventListener( 'click',  function ( ) {
    dlog.close( )
    info.innerText = 'Modal Dialog Confirmed' } )
} ) ( )

</script>
```

5 Save the HTML document, then open it in your browser and click the button to see the modal dialog

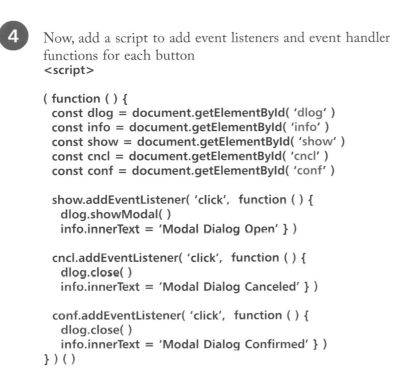

Hot tip

Although the **onclick** and **addEventListener** techniques are both correct, there are advantages in preferring event listeners. It cleanly separates script from HTML code and, unlike **onclick**, it allows you to have multiple listeners for the same event.

Beware

In Firefox you may have to open **about:config** and set **dom.dialog_ element** to **enabled** to see the modal dialog.

Provide Widgets

Web widgets can be included on a web page to provide additional functionality for the user. These are small applications that provide access to resources from another website. There are many web widgets that offer facilities such as social media buttons, the latest news, current weather, user comments, and page customization.

You do not need to write your own web widgets, as code is freely available online by searching for "web widgets". For example, you could provide an exchange rate widget on a web page offering a property for sale that might attract international interest.

widget.html

widget-resort.jpg

 Create an HTML document

 Within the body section, add a heading
<h1>Apartment for Sale - $50,000</h1>

 Next, add a main section that contains an image
<main>
<img src="widget-resort.jpg"
width="300" height="225" alt="Resort">
</main>

4 Now, add an aside section that will contain a widget
<aside>
<!-- Widget code to be inserted here. -->
</aside>

5 Then, add a style sheet to position the main and aside sections side-by-side at the left of the page
<style>
main { position : absolute ;
top : 50px ; left : 10px ; width : 320px ; }
aside { position : absolute ;
top : 80px ; left : 340px ; width : 200px ; }
</style>

6 Open your web browser and search for "exchange rate widget", then choose one of the many results – for instance, you might select **www.exchangeratewidget.com**

7 From the website's form, select the currencies you want to provide and any other options you prefer

Choose a currency in the left pane then click the right arrow button to add it to the right pane as a selection to appear in the widget.

8 Copy the code provided for your selections, then paste it into the aside section of your web page

```
<!-- Exchange Rates Script - EXCHANGERATEWIDGET.COM -->
<div style="width:198px;border:3px solid #FF6600;border-radius:5px;"text-align:left;"><div
style="text-align:left;background-color:#FF6600;width:100%;border-bottom:0px;height:16px; font-
size:12px;font-weight:bold;padding:5px 0px;"><span style="margin-left:2px;background-
image:url(//www.exchangeratewidget.com/flag.png); background-position: 0 -1232px; width:100%;
height:9px; background-repeat:no-repeat;padding-left:5px;"><a
href="https://www.exchangeratewidget.com/" style="color:#FFFFFF; text-decoration:none;padding
```

9 Save the HTML document alongside the image on a web server, then open it in your browser to see the widget

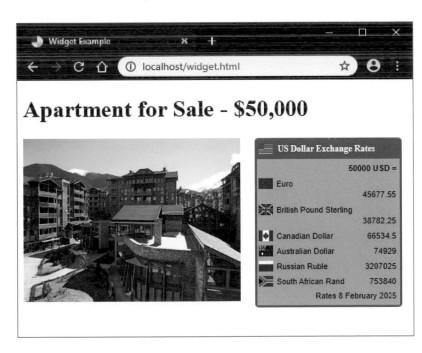

This example will not grab the widget unless it is located on a web server to access the remote resource.

Summary

- The **<img>** tag places an image on the web page, and should preferably always include **src, width, height,** and **alt** attributes.

- The **<map> </map>** tags enclose **<area>** elements, to define the areas of an image map, and must include a **name** attribute.

- The **<figure>** and **<figcaption>** tags can be used to embed a captioned reference image within an HTML document.

- A **<picture>** element can contain several **<source>** tags to provide a variety of image sizes for different screen widths.

- The **<object> </object>** tags can be used to embed external resources within an HTML document.

- External resources can be embedded into an inline frame with an **<iframe>** element that sets the size of a display area.

- External audio resources can be embedded into an HTML document using **<audio> </audio>** tags, and external video resources can be embedded using **<video> </video>** tags.

- A **<video>** element can contain a nested **<track>** element to specify a subtitle file in the WebVTT format.

- The **<progress>** element can be used to provide a visual indicator of media playback.

- The **<template>** element can designate a group of elements that can be cloned to dynamically write content.

- A template can include **<slot> </slot>** elements whose content can differ in each instance of the template.

- The **<dialog> </dialog>** tags create a modal dialog on a layer above all other page content.

- The **<button> </button>** tags create a clickable button.

- Web widgets can be included on a web page by pasting freely-available code into an HTML element.

6 Create a Local Domain

Install Abyss

It is useful to have a web server installed on your system for web page development. There are several free web servers available, such as the Abyss Web Server X1 from Aprelium.

1 Launch your web browser and visit **aprelium.com**

2 Click the **Downloads** menu item, then choose **Free Download** and download the installer for your system

Hot tip

There are versions of Abyss for Windows, macOS, and Linux.

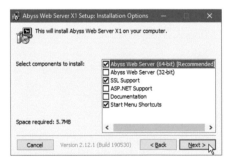

Hot tip

On a Windows system, double-click on the downloaded **.exe** file to run the installer.

3 Run the installer, accept the license agreement, and choose options – such as **64-bit** version, **SSL Support**, and **Start Menu Shortcuts**

4 Accept the suggested installation location and click the **Install** button

5 Choose how you prefer the web server to be started – such as **Install as a Windows Service** to continuously run in the background automatically

6 Click the **OK** button to open the Abyss Web Server Console

7 Select your language (**English**), then enter a memorable login name and password – you will need these later!

8 Click the **OK** button to save your credentials, then click the next **OK** button to see that the web server is running

9 Now, type **localhost** into your browser's address field, then hit **Return** to see a default index web page appear

Install Python

A web server can run server-side scripts that respond to requests made from a web browser. The most popular server-side scripting languages include PHP, ASP.NET, Ruby, Perl, and Python.

In order to enable server-side scripting for the Abyss Web Server installed on your PC, you must first install the interpreter for a server-side scripting language – such as Python.

 Launch your web browser and visit **activestate.com**

 Sign in (or create an account and sign in), then select **Featured Projects & Languages** and download the latest Python build installer for your system

Hot tip

Python is available for free and there is no charge for creating an account at ActiveState.

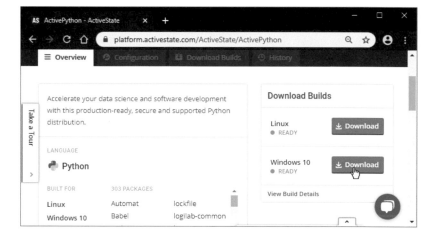

 Run the installer, then click **Next** to begin setup

 Click **Next** to accept the license agreement

Don't forget

The suggested installation folder name may include a version number, but you can change it to **C:\Python**.

Click **Next** to accept the suggested installation location and all options

6 Now, click the **Install** button on the next dialog to begin the installation process – sit back, this will take a while

Installation is lengthy because the installer dynamically compiles C source code as it runs.

7 When installation has completed, click the **Finish** button to close the installer

You can find the Command Prompt launcher in the Windows System folder on the Windows 10 Start menu.

8 Now, open a Command Prompt or Terminal window, then type **python** and hit **Return** to open the Python Console – where you can interact with the interpreter

9 At the Python Console prompt, precisely type this line of Python code and hit **Return** to see the location of the interpreter "python.exe" on your system
import sys ; print(sys.executable)

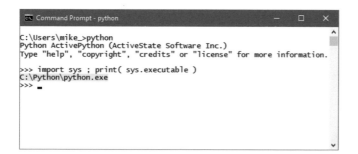

```
C:\Users\mike_>python
Python ActivePython (ActiveState Software Inc.)
Type "help", "copyright", "credits" or "license" for more information.

>>> import sys ; print( sys.executable )
C:\Python\python.exe
>>>
```

Note the interpreter location, as you will need it to configure Abyss for Python scripting.

127

Configure Abyss

In order for the Abyss Web Server to execute Python server-side scripts it must be configured to know the location on your system of the Python interpreter (**python.exe**) and to recognize that files with the file extension of **.py** are Python scripts.

1 With the Abyss Web Server running, launch your web browser and type **localhost:9999** into the address field then hit **Return** to open the Abyss Web Server Console

2 Enter your credentials, then click the **Configure** button

3 Now, click the **Scripting Parameters** icon

Hot tip

The domain name **localhost** is an alias for the IP (Internet Protocol) address **127.0.0.1** – so you could enter **127.0.0.1:9999** to open the Abyss Web Server Console.

4 A "Scripting Parameters" page will now appear – check the **Enable Scripts Execution** box

5 Now, click the **Add** button in the Interpreters table

6 An "Interpreters" page will now appear – set the Interface field to **CGI/ISAPI**

7 Now, click the **Browse** button in the Interpreter field, and go to the Python folder and select the **python.exe** file

8 Check the **Use the associated extensions to automatically update the Script Paths** box

9 Now, click the **Add** button in the Associated Extensions field and enter **py** in the Extension field

Hot tip

CGI (Common Gateway Interface) and ISAPI (Internet Server Application Programming Interface) allow data to be transferred between the web server and a script interpreter.

Don't forget

Your selections should look similar to those shown here.

129

10 Click the **OK** button, and click **OK** again, then click the **Restart** button to apply the new configuration

Echo Script

Having configured the Abyss Web Server for the Python interpreter, and for script files with the **.py** file extension, you can now create a Python script to be executed in response to a request from your web browser.

Many requests send data as key=value pairs to the web server, so a Python script could simply echo the pairs in an HTML response that places the keys and values in a table.

echo.py

1 Open a plain text editor, such as Windows' Notepad app, then type an instruction at the beginning of the first line to use a special Python module for server-side scripting
import cgi

2 Next, assign all key=value pairs to a variable, using a function supplied by the special Python module
form = cgi.FieldStorage( )

3 Now, use the Python built-in **print( )** function to write the HTML headers that will be sent to the browser
print('Content-Type:text/html; charset=utf-8')

4 The headers must be separated from content, so add this line to write two carriage return and newline characters
print('\r\n\r\n')

5 Start writing the HTML document with a type declaration, title, and style sheet
print('''<!DOCTYPE HTML>
<html><title>Web Server Response</title>
<style>tr,th,td{border:2px solid Gray}</style>''')

6 Next, begin a table with two header cells
print('<table style="width:500px"><tr><th>Key<th>Value')

Indentation is used in Python to group statements, instead of curly brackets, so these three lines must be indented alike.

7 Then, add a loop to write any keys and values in table cells
for i in form.keys( ) :
** key = str(i)**
** val = str(form.getvalue(key))**
** print('<tr><td>' + key + '<td>' + val)**

8 Finally, add these lines to complete the table, to display the server icon, and to complete the HTML document
print('''</tr></table>
<img src="pwrabyss.gif"></html>''')

```
echo.py - Notepad                              —     □    ×
File  Edit  Format  View  Help
import cgi

form = cgi.FieldStorage( )

print('Content-Type:text/html; charset=utf-8')

print('\r\n\r\n' )

print('''<!DOCTYPE HTML>
<html><title>Web Server Response</title>
<style>tr,th,td{border:2px solid Gray}</style>''')

print( '<table style="width:500px"><tr><th>Key<th>Value' )

for i in form.keys( ):
  key = str( i )
  val = str( form.getvalue( key ) )
  print( '<tr><td>' + key + '<td>' + val )

print('''</tr></table>
<img src="pwrabyss.gif"> </html>''')
```

Hot tip

You can discover more about Python scripting with the companion book in this series: Python in easy steps.

9 Ensure that your script looks like the screenshot above

10 Save the file as **echo.py** in the **htdocs** folder of the Abyss Web Server location on your system – typically on Windows this is **C:\Abyss Web Server\htdocs**

11 With the Abyss Web Server running, open your browser then type **localhost/echo.py** into the address field and hit **Return** to see the script respond with an empty table

Web Server Response   ×   +   —   □   ×

← → C ⌂   ⓘ localhost/echo.py   ☆   😀   ⋮

Key	Value

Powered by
Abyss
Web Server

Hot tip

The **echo.py** script is used throughout Chapter 7 to demonstrate HTML form submission to the Abyss Web Server.

Test Environment

Now, with the Abyss Web Server installed and configured to execute the Python script on pages 130-131, the environment can be tested by sending data from the browser to the server.

Data can be sent to the server by appending a "query string" to the URL of the Python script. This begins with a **?** question mark separator followed by a key=value pair – for example, to send a single pair with **http://localhost/echo.py?Forename=Mike**.

Multiple key=value pairs can be sent to the server with each pair separated by an **&** ampersand character in the query string, such as **http://localhost/echo.py?Forename=Mike&Surname=McGrath**.

The data can be sent by entering the URL and query string directly into the browser's address field. It can also be sent from a hyperlink that targets the URL and query string, or from JavaScript that sets location to the URL and query string.

query-string.html

 Create an HTML document

 Next, within the body section, add a hyperlink to send data to the Python script
<a href="http://localhost/echo.py?CSS=Style">Styling Web Pages</a>

 Now, add a button in the body section
<button id="sender">Scripting Web Pages</button>

 Finally, add a script to send data to the Python script when the user clicks the button
<script>

```
( function ( ) {
  const sender = document.getElementById( 'sender' )
  sender.addEventListener( 'click' , function ( ) {
  location='http://localhost/echo.py?JavaScript=Function' } )
} ) ( )
```

</script>

 Save the HTML document in the server's **htdocs** folder, then open it in your browser and use the browser's address field, hyperlink, or button to send data

...cont'd

The Python script is performing as expected. Add more key=value pairs to the query string (each pair separated by an **&**) to see the response add further rows to the table

133

Summary

- The free Abyss Web Server X1 can be installed on your own PC for web development.

- Abyss can be started manually or set to continuously run in the background automatically.

- When Abyss is running it provides a domain named **localhost** that is an alias for the IP address **127.0.0.1**.

- A web server can call upon an interpreter to execute server-side scripts that respond to requests from a web browser.

- The most popular server-side scripting languages include PHP, ASP.NET, Ruby, Perl, and Python.

- When Python is installed you can directly interact with its interpreter (**python.exe**) via the Python Console.

- Abyss must be configured to know the location of the Python interpreter (**python.exe**) and to recognize that files with the file extension of **.py** are Python scripts.

- The URL **localhost:9999** and user credentials are needed to open the Abyss Web Server Console.

- The CGI/ISAPI interfaces allow data to be transferred between the web server and a script interpreter.

- The Abyss Web Server's location on your system contains a **htdocs** folder that is recognized by the **localhost** domain.

- HTML documents, server-side scripts, and other resource files should be placed in the Abyss **htdocs** folder.

- Data can be sent to a web server by appending a query string to a requested URL.

- Query strings begin with a **?** question mark separator.

- Query strings contain one or more key=value pairs, with each pair separated by an **&** ampersand character.

- A Python server-script can echo data sent to it from a browser request within a web server response.

7 Produce Input Forms

Submit Text

Web page forms are built from HTML elements that send data to a web server. Each element includes a **name** attribute and a **value** attribute so the data assigned to these attributes can be processed as key=value pairs. For example, where an element's **name** attribute is assigned "Brand" and its **value** attribute is assigned "Ford", the key=value pair represents the data as Brand=Ford.

Form elements are enclosed between **<form>** **</form>** tags. Each opening **<form>** tag should include a **method** attribute, specifying which HTTP method is to be used to submit the form, and an **action** attribute specifying the URL of a server-side script that is to be used to process the submitted data.

The **method** attribute can be assigned values of "GET" or "POST". Submission via the preferred **GET** method appends the data to the URL, whereas submission via the **POST** method encodes the data differently and can be used when the **GET** method fails.

Data sent by the GET method is attached to the request as a query string, so may be visible in the browser's address field in the web server's response. You can submit by the POST method to prevent this for sensitive data.

Typically, an HTML form will have a "Submit" button that the user clicks to send the data for processing. This is created by assigning the value "submit" to a **type** attribute of an **<input>** tag. Additionally, this tag may include **name** and **value** attributes to submit data assigned to them as a name=value pair.

An HTML form can provide text boxes where the user can input data for submission. These are created by assigning the value "text" to the **type** attribute of an **<input>** tag, and a name to its **name** attribute. The data in the text box is sent as the value associated with the text box name as a key=value pair. Optionally, the **<input>** tag can include a **value** attribute to specify a default value. A text box for the input of a password is created by assigning the value "password" to the **type** attribute of an **<input>** tag. This functions like any other text box, but it does not display its contents as readable text. Both text and password **<input>** elements can optionally include these other attributes:

- **size** – the width of the text box in average character widths.
- **minlength** and **maxlength** – permissible number of characters.
- **min** and **max** – permissible range of numeric values.
- **placeholder** – provides a data entry hint to the user.
- **readonly** – the default value in the text box cannot be changed.
- **disabled** – the text box is grayed out and will not be submitted.

① Create an HTML document containing a form to send data to a server-side script using the GET method

```
<form method="GET" action="http://localhost/echo.py" >
<!-- Form components to go here. -->
</form>
```

textbox.html

② Now, in the form element, insert text inputs and a submission button – whose value will appear on the button

```
<dl>
<dt>User Name:
<dd><input type="text" name="Name">
<dt>Password:
<dd><input type="password" name="Pwd">
<dt>Zip Code:
<dd><input type="text" name="Zip"
        size="5" maxlength="5">
</dl>
<input type="submit" name="Form" value="Sender">
```

The examples in this chapter send form data to the server-side script created on page 130.

③ Save the HTML document in the Abyss **htdocs** folder, then open it in your browser and submit some data

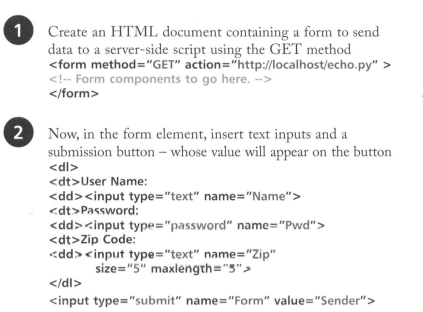

Notice that the data in the response is not necessarily in the same order as the form input elements.

Input Types

An HTML form **<input>** tag can enforce its completion by including the **required** keyword. It can also control what the user is permitted to submit by the value assigned to its **type** attribute. Many of the input types listed in the table below prohibit submission of the form if the user enters a value that is not permitted and issue an error notice. Some also provide special controls that allow the user to easily select a permitted value.

Type	Permitted input
text	String of text
password	String of text (obscured by browser)
url	Valid URL protocol and domain address
email	Valid email address
date	Date in mm/dd/yyyy format
month	Month and year
week	Week number and year
time	Time in HH:MM format
datetime-local	Date and time as mm-dd-yy HH:MM
number	Numeric integer value
range	Numeric integer value (slider)
color	Color in #RRGGBB hexadecimal format
file	File path address (browse)

input-types.html

 Create an HTML document containing a form with a submit button
```
<form method="GET" action="http://localhost/echo.py" >
<!-- Input elements to go here. -->
<p><input type="submit" value="Submit Form"></p>
</form>
```

2 Now, in the form, insert four controlling input elements
```
Color: <input type="color" name="color">
Range:
<input type="range" name="range" min="1" max="10">
Time: <input type="time" name="time"> <br>
URL:
<input type="url" name="url" size="54" required >
```

3 Save the HTML document in the Abyss **htdocs** folder, then try to submit the form to see it fail

4 Enter a valid URL into the **url** type input field

5 Now, enter permitted values using the special controls for the other inputs, then submit the form successfully

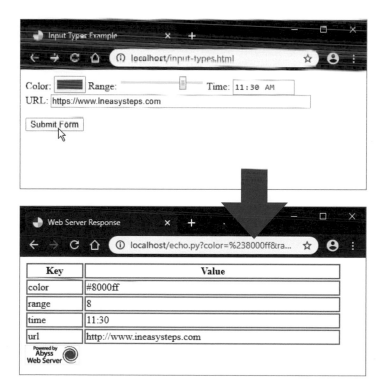

Notice that special characters are converted to percentage Unicode values in the query string. Here the **#** symbol is converted to its Unicode value **%23**.

Text Areas

An HTML form can provide a multi-line text field where the user can input data for submission to the web server for processing. These are created by **<textarea> </textarea>** tags that may enclose default text content.

The **<textarea>** tag should include a **name** attribute that will be associated with the element's content upon submission as a key=value pair. Additionally, this tag must include a **rows** attribute, to specify the number of visible text lines, and a **cols** attribute to specify the field width in average character widths. Optionally, it may also include a **readonly** attribute to prevent the user editing its content.

When submitting large bodies of text you must be aware of some limitations of the **GET** method. This varies by browser, but may only allow the URL to append up to around 200 characters. The **POST** method provides much larger capacity as the text is sent as "Form Data" along with the HTTP header, not simply appended to the URL:

textarea.html

Unlike a text **<input>** element, the **<textarea>** element has no **value** attribute – as its content is treated as its value.

1 Create an HTML document with a form element containing a submit button to send form data by the POST method
<form method="POST" action="http://localhost/echo.py" >

```
<!-- Text area element to go here. -->
```

<p><input type="submit" value="Submit Form"></p>

</form>

2 Now, in the form element, insert a text input area that has 10 rows and is 65 average character widths wide
<textarea name="The Future Web"
**   rows="8" cols="70">**
</textarea>

3 Save the HTML document in the Abyss **htdocs** folder, then open the web page in your browser, enter some data, and submit the form

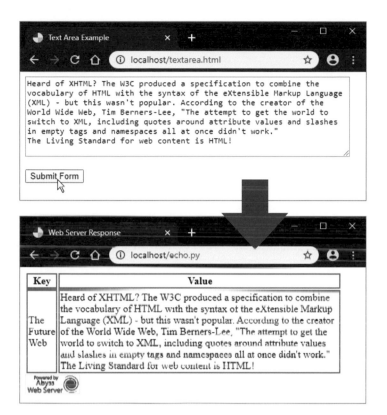

 The text is not appended to the URL, so examine the response headers to see it has been sent as "Form Data"

Check Boxes

An HTML form can provide a visual checkbox "on/off" switch that the user can toggle to include or exclude its associated data for submission to the server. When the box is checked, the switch is set to "on" and its key=value pair will be submitted.

A checkbox is created by assigning the value "checkbox" to the **type** attribute of an **<input>** tag. This tag must also include a **name** attribute and a **value** attribute to specify the key=value pair values. Optionally, this tag may also include a boolean **checked** attribute to set the initial state of the switch to "on" – so a check mark will automatically appear in the checkbox. Checkbox names may be individually unique, or several checkboxes can share a common name to allow the user to select multiple values for the same named property. In this case, the selected values are returned by the server as a comma-separated list where key=value,value,value.

A "radio button" is similar to a checkbox but is created by assigning the value "radio" to the **type** attribute of an **<input>** tag. Unlike checkboxes, radio buttons that share a common name are mutually exclusive, so when one radio button is selected, all others in that group are automatically switched off.

Multiple checkboxes and radio buttons can be visually grouped by surrounding their **<input>** elements with **<fieldset> </fieldset>** tags. These may also contain **<legend> </legend>** tags to state a common group name:

Hot tip

The **checked** attribute need have no assigned value – its presence sets the switch to "on" and its absence leaves the switch in its "off" state.

HTML

checkbox.html

 1 Create an HTML document with a form element containing a submit button
<form method="GET" action="http://localhost/echo.py" >

`<!-- Checkbox and radio buttons to go here. -->`

<p><input type="submit"></p>
</form>

2 Now, in the form element, insert a paragraph containing two checkboxes
<p>Send details
<input type="checkbox" name="Send" value="Details">
Send prices
<input type="checkbox" name="Send" value="Prices">
</p>

3 Next, in the form element, insert a fieldset with a legend

```
<fieldset>
<legend>What kind of language is HTML?</legend>

<!-- Radio buttons to go here. -->

</fieldset>
```

The **<fieldset>** element only groups the related elements it encloses for visual presentation – it does not associate them programmatically.

4 Then, in the fieldset, insert radio buttons with one selected

```
Scripting <input type="radio"
       name="Definition" value="Scripting"> <br>
Markup <input type="radio"
       name="Definition" value="Markup"> <br>
Programming <input type="radio"
name="Definition" value="Programming" checked>
```

5 Save the HTML document in the Abyss **htdocs** folder, then open the web page, check both checkboxes, select the correct radio button answer, and submit the form

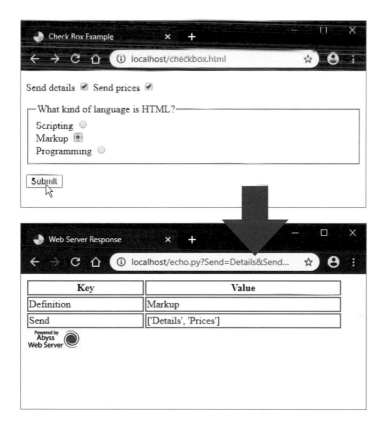

Check only one checkbox and submit the form to see only the checkbox's name and value are sent to the web server.

Hide Data

An HTML form can provide hidden elements that create no visible controls but allow additional data to be submitted to the server. Hidden form data is created by assigning the value "hidden" to the **type** attribute of an **<input>** tag. This tag must also include a **name** attribute, and may include a **value** attribute to specify static data that will be submitted as a name=value pair. Optionally, the **<input>** tag may include an **id** attribute and omit the **value** attribute so its value can be specified by script.

Hidden form data can also be used to perform a calculation and dynamically display the result in an **<output> </output>** element. The **<output>** tag must include an **id** attribute and a **for** attribute for reference in script. The **for** attribute can specify multiple element identities as a space separated list. These can be used in an assignment to the **<form>** tag's **oninput** attribute to perform a calculation whose result will appear in the **<output> </output>** element – but will not be submitted to the server:

hidden.html

 Create an HTML document containing an image displaying an item with sale price details
```
<img src="hidden-sale.png"
        width="200" height="120" alt="Sale">
```

 Next, insert a form element
```
<form method="GET" action="http://localhost/echo.py" >

<!-- Hidden data, input, and output to go here -->

<input type="submit" name="Offer"
        value="Buy Teddy Bears">
</form>
```

 Within the form element, insert a visible input element for user-entered data
```
Qty (60 Available) <input type="number" id="qty"
name="Quantity" min="1" max="60">
```

4 Now, in the form element, insert an invisible element for hidden data and an element to display a calculated result
```
<input type="hidden" id="price"
        name="Unit Price" value="24.99">
<output name="sum" for="qty price"></output> <br>
```

5 Then, insert another attribute in the **<form>** tag
oninput="sum.value=multiply(qty, price)"

6 Add a script to perform the calculation
<script>

```
function multiply( q, p ) {

  let result=parseFloat( q.value ) * parseFloat( p.value )

  if ( isNaN( result ) || result < 1 ) return ' '
  else return 'Total: $' + result.toFixed( 2 )
}
```

</script>

JavaScript is case-sensitive so you must use the correct case when copying script examples.

7 Save the HTML document in the Abyss **htdocs** folder, then open the web page in your browser, enter data, and submit the form

The hidden form **<input>** data gets submitted to the server, but the **<output>** element merely displays the result of the calculation.

Upload Files

An HTML form can provide a file selection facility, which calls upon the operating system's "Choose File" dialog, to allow the user to browse their local file system and select a file.

A file selection facility is created by assigning the value "file" to the **type** attribute of an **<input>** tag and a name to its **name** attribute. This element produces a text field and a "Browse" button to launch the Choose File dialog. After a file has been selected, its full path appears in the text field. When the form is submitted, the element name and the selected file's name are sent to the web server as a name=value pair.

Where a selected file is to be uploaded to the web server, the **<form>** tag must include an **enctype** attribute specifying the encoding type as "multipart/form-data". Also, its **method** attribute must specify the POST method – because Form Data cannot be appended to a URL using the GET method:

upload.html

 Create an HTML document with a form element containing a submit button to send form data by the POST method and specify the encoding type for Form Data

```
<form  method="POST"
       action="http://localhost/upload.py"
       enctype="multipart/form-data" >

<!-- File input element to go here. -->

<input type="submit">
</form>
```

 In the form element, insert a file element and a line break
```
<input type="file" name="Upload" size="70" > <br>
```

 Save the HTML document in the Abyss **htdocs** folder, then open the web page in your browser, select a file on your system, and submit the form

 Look in the Abyss **htdocs** folder to see a copy of the selected file is now placed there

This example uses a Python script named **upload.py** placed in the Abyss **htdocs** folder. This script is provided in the download archive for this book, which is freely available from **www.ineasysteps. com/resource-centre/ downloads**

Push Buttons

An HTML form can provide push buttons for scripting purposes. When the user pushes a button, a "click event" occurs to which a script function can respond. This allows the user to dynamically interact with the form, and can be used to set attribute values. When a script designates a function to be called, whenever a button gets pushed it is said to attach a "behavior" to that button. A push button is created by specifying a "button" value to the **type** attribute of an **<input>** tag, and should also include an **id** attribute so the script can easily identify that element. Text assigned to the button's **value** attribute will appear on the face of the button.

Additionally, any HTML form can be returned to its original state by pushing a reset button that is created by specifying a "reset" value to the **type** attribute of an **<input>** tag:

button.html

1 Create an HTML document with a form element containing a reset button, a push button, and a submit button
<form method="GET" action="http://localhost/echo.py" >

<!-- Fieldset to go here. -->

<input type="reset" value="Reset Form">
<input type="button" value="Choose For Me" id="btn">
<input type="submit" value="Submit Form">

</form>

2 Within the form element, insert a fieldset containing a legend and a checkbox group
<fieldset>

<legend>Pizza Toppings</legend>

**<input id="pepperoni" type="checkbox"
  name="Toppings" value="Pepperoni"> Pepperoni |**

**<input id="mushroom" type="checkbox"
  name="Toppings" value="Mushroom"> Mushroom |**

**<input id="bbqsauce" type="checkbox"
  name="Toppings" value="BBQ Sauce"> BBQ Sauce**

</fieldset>

3 Add a script that attaches a behavior to the push button

```
<script>
( function ( ) {

const pep = document.getElementById( 'pepperoni ' )
const btn = document.getElementById( 'btn ' )

btn.addEventListener( 'click' , function ( ) {
                        pep.checked = true } )
} ) ( )
</script>
```

Don't forget

The mere presence of a boolean **checked** attribute in an HTML element checks the box, but in script the box's **checked** property needs to be assigned a **true** value to check the box.

4 Save the HTML document in the Abyss **htdocs** folder, then open the web page in your browser, and push the button to check a box

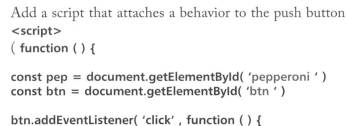

5 Now, push the reset button to clear the form, then check the other two boxes and submit the form

Key	Value
Toppings	['Mushroom', 'BBQ Sauce']

Image Buttons

An HTML form can use an image button to submit the form, in place of a regular submit button. An image button is created by specifying an "image" value to the **type** attribute of an **<input>** tag and including an **alt** attribute. When a form is submitted by an image button, the XY coordinates of the point at which the click occurred are automatically submitted as key=value pairs along with the rest of the form data.

Additionally, a regular **<img>** tag can be used as an image button by attaching a behavior with script. Where the behavior is to submit a form, the script function can usefully incorporate validation – for example, to ensure a user-entered email address is in the expected format:

ibutton.html

 Create an HTML document with a form element containing a text input field, which both have an identity for scripting

```
<form id="my-form"
        method="GET" action="http://localhost/echo.py" >

Please Supply Your Email Address:
<input id="adr"
        name="Address" type="text" size="45"> <br>

<!-- Image Buttons to go here. -->

</form>
```

Note that the image button that will perform validation is given an identity so the script can attach a behavior to it.

 Next, in the form element, insert an image button that will simply submit the form

```
<input type="image"
        src="ibutton.png"
        alt="Submit Button"
        title="Click to submit form">
```

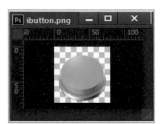

ibutton-png

Now, in the form element, insert an image button that will perform validation then submit the form

```
<img    id="btn"
        src="ibutton.png"
        alt="Submit Button"
        title="Click to submit with JavaScript validation">
```

150

4 Add a script that attaches a behavior to an image button

```
<script>
( function ( ) {

const btn = document.getElementById( 'btn' )
btn.addEventListener( 'click' , function ( ) {

  const myForm =  document.getElementById( 'myForm' )
  const pattern =
  /^([a-zA-Z0-9_.-])+@([a-zA-Z0-9_.-])+\.([a-zA-Z])+([a-zA-Z])+/
  let adr = document.getElementById( 'adr' ).value
  if( ! pattern.test( adr ) ) alert( 'Invalid Email Address' )
  else myForm.submit( )
 } )

} ) ( )
</script>
```

Beware

The script in this example checks the input text against a regular expression pattern that describes the format of any valid email address. The pattern must appear on a single line – exactly as it is listed here.

5 Save the HTML document in the Abyss htdocs folder, then open it in your browser, enter an incomplete email address, and submit the form using each button

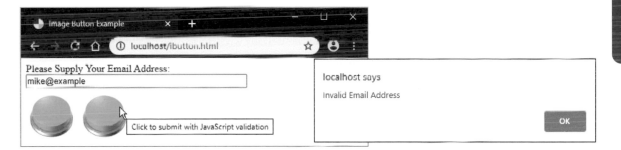

6 When validation fails using the button with scripted behavior, correct the email address then click the validating button again to submit the form successfully

Hot tip

Submit the form with the unscripted image button to also see coordinates in the web server response.

Add Logos

HTML can create push buttons that display small "logo" images using **<button> </button>** tags. These tags can then enclose an **<img>** element specifying the URL of the logo image, and text that will appear on the face of the button.

Each **<button>** tag should include a **type** attribute to specify whether the button is simply a scripting "button" type, a "submit" form type, or a "reset" form type. Scripting buttons can include an **onclick** attribute in the **<button>** tag to specify the function to be called when the button gets clicked, or directly specify a snippet of script to execute:

logo.html

 Create an HTML document with a form element containing a fieldset with a legend and a text input field
<form method="GET" action="http://localhost/echo.py">

<fieldset>

<legend>Favorite Color</legend>
<input type="text" name="Color">

<!-- Logo Buttons to go here. -->

</fieldset>

</form>

 In the fieldset, insert a scripting logo button specifying a snippet of script to execute when that button gets clicked
<button type="button"
onclick="alert('Enter your favorite color in the text box')">

<!-- Logo Image and Face Text to go here. -->

</button>

 Now, within the button element, insert an image element and text that will appear on the face of the button
<img src="logo-help.png" alt="Help">Help

4 Next, add a button element to submit the form
<button type="submit">
<img src="logo-submit.png" alt="Submit">
Submit</button>

Finally, add a button element to reset the form
<button type="reset">
<img src="logo-reset.png" alt="Reset">Reset</button>

 Save the HTML document in the Abyss **htdocs** folder, then open it in your browser and click the "Help" button

You can specify a default value for a text input to the **value** attribute of its **<input>** tag.

Enter a color in the text box, then click the reset logo button to clear the text box

Enter a color in the text box again, then submit the form

Select Options

An HTML form can provide a select option list, from which the user can select one option to include its associated data for submission to the server.

A select option list is created using **<select>** **</select>** tags. The opening **<select>** tag must include a **name** attribute specifying a list name. The **<select>** element encloses **<option>** **</option>** tags that define each option. Each opening **<option>** tag must include a **value** attribute specifying an option value. When the form is submitted, the list name and the selected option value are sent to the server as a name=value pair.

Optionally, one **<option>** tag may also include a boolean **selected** attribute to automatically select that option, and the **<option>** elements may be grouped by enclosure in **<optgroup>** **</optgroup>** tags. The opening **<optgroup>** tag may specify an option group name to a **label** attribute.

A select option list will normally appear as a single-line dropdown list unless a **size** attribute is included in the **<select>** tag to specify the number of rows to be visible:

select.html

 Create an HTML document with a form element containing a submit button
<form method="GET" action="http://localhost/echo.py" >

<!-- Select option lists to go here. -->

<p>
<input type="submit">
</p>

</form>

2 Now, in the form element, insert a fixed height select option list with one option automatically selected
<select name="HTML List Type Selector One" size="4">

<optgroup label="List Type 1">
<option value="UL">Unordered List</option>
<option value="OL" selected>Ordered List</option>
<option value="DL">Description List</option>
</optgroup>

</select>

3 Next, in the form element, insert a dropdown select option list with one option automatically selected

```
<select name="HTML List Type Selector Two">

<optgroup label="List Type 2">
<option value="UL">Unordered List</option>
<option value="OL">Ordered List</option>
<option value="DL" selected>Description List</option>
</optgroup>

</select>
```

4 Save the HTML document in the Abyss **htdocs** folder, then open it in your browser

5 Open the dropdown list and submit the form to see the default option values get submitted

Always include a **selected** attribute to automatically select one option in each option list – to provide a default choice.

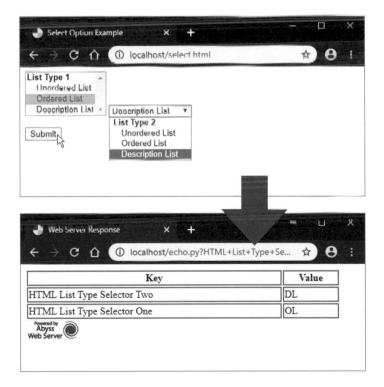

Datalist Options

A simple "autocomplete" feature can be provided for a text **<input>** using a **<datalist>** **</datalist>** element to enclose a number of pre-defined **<option>** values. The user may choose any one of the options, or enter text directly into the input field. In order to associate the **<input>** field with the list, the **<datalist>** tag must include an **id** attribute to specify a list name. The same name must then be specified to a **list** attribute within the **<input>** tag to create the association. The **<input>** tag must also include a name attribute to send to the server as usual.

datalist.html

 Create an HTML document with a form element containing a submit button
<form method="GET" action="http://localhost/echo.py"**>**

<!-- Form data list and input field to go here. -->

<p><input type="submit" **value=**"Submit Form"**></p>**
</form>

The key=value pair submitted to the server is the specified list name and the input value selected from the options or entered directly by the user.

 Next, insert a data list of pre-defined options with a specified **id** name
<datalist id="browsers"**>**

<option value="Google Chrome"**>**
<option value="Firefox"**>**
<option value="Internet Explorer"**>**
<option value="Opera"**>**
<option value="Safari"**>**
<option value="Microsoft Edge"**>**

</datalist>

 Now, insert a label that contains text and an input field that is associated with the data list above
<label>
Choose your browser from this list:
<input list="browsers" **name=**"myBrowser"**>**
</label>

④ Save the HTML document in the Abyss **htdocs** folder, then open the web page in your browser

5 Select the input field to see the pre-defined options appear in a dropdown list

You may need to double-click the input field to override your browser's own autocomplete suggestions.

6 Select any option from the dropdown list, or type your own text into the input field to create a value

7 Submit the form to send the input field name and your chosen value to the server

Label Controls

Text that is to be associated with an HTML form control can be enclosed between **<label> </label>** tags. The opening **<label>** tag can include a **for** attribute to specify the value assigned to the control's **id** attribute to make the association.

Alternatively, the **<label>** element can simply enclose both the text and the control element to make the association. This allows styling to be applied to the entire label – including the text and control. Often this is useful to distinguish the control associated with particular text.

Additionally, each form control element may include a **tabindex** attribute to specify its tabbing order within the document as a unique value between 0 and 32,767. Using the tab key, the user can then navigate through the document starting at the lowest **tabindex** value and proceeding through successively higher values:

label.html

 1 Create an HTML document with a form element containing a fieldset with a legend
```
<form method="GET" action="http://localhost/echo.py">
<fieldset> <legend>Toolbox</legend>
<!-- Form Controls to go here. -->
</fieldset>
</form>
```

Hot tip

A form "control" is any **<input>**,**<button>**, or **<textarea>** element. A **tabindex** attribute can be included in these tags and also in any **<a>**, **<area>**, **<object>**, or **<select>** tag.

2 Now, in the fieldset, insert labels that each contain text and a checkbox with a specified tab order
```
<label>Hammer
<input type="checkbox" name="Toolbox"
        value="Hammer" tabindex="2" checked></label>
<label>Screwdriver
<input type="checkbox" name="Toolbox"
        value="Screwdriver" tabindex="3" ></label>
<label>Wrench
<input type="checkbox" name="Toolbox"
        value="Wrench" tabindex="4" checked></label>
<label>Drill
<input type="checkbox" name="Toolbox"
        value="Drill" tabindex="5"></label>
<label>Saw
<input type="checkbox" name="Toolbox"
        value="Saw" tabindex="6" checked></label>
```

 3 Next, insert a logo submit button – in first tab place
```
<button type="submit" tabindex="1">
<img src="label-tools.png" alt="Tools">Submit</button>
```

4 Save the HTML document in the Abyss **htdocs** folder, then open it in your browser to see the text-control association is unclear

5 Edit the HTML document to add a class attribute to each alternate label tag for styling purposes
```
<label class="hilite">
```

6 Add a style sheet with a rule to distinguish the labels
```
<style>
label.hilite { background : Red ; color : White ; }
</style>
```

7 Save the HTML document again, then open the web page to see that the text-control association is now clear. Use the tab key to move between controls and the space bar to select checkboxes, then submit the form

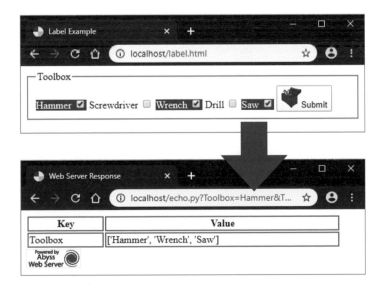

Summary

- HTML forms submit data to the web server as key=value pairs for processing by a specified server-side script.

- All form component elements are enclosed between **<form>** **</form>** tags, which must include an **action** attribute, to specify the URL of the processing script, and a **method** attribute to specify the submission method as **GET** or **POST**.

- Each **<input>** tag's **type** attribute specifies its component type, such as "text", "password", "checkbox", "radio", etc.

- An **<input>** tag can include **name** and **value** attributes to specify data for submission as a key=value pair.

- An **<input>** tag can enforce its completion by including the **required** keyword, but many automatically prohibit submission if the user enters a value that is not permitted.

- A multi-line text field is created by **<textarea>** **</textarea>** tags that require **rows** and **cols** attributes to specify its size.

- Radio button and checkbox inputs only submit their **name** and **value** attribute data if they are checked.

- Forms can contain "hidden" elements that allow static or script-generated data to be submitted to the server for processing.

- When a form is to upload files, the **<form>** tag must include an **enctype** attribute specifying encoding as "multipart/form-data".

- A form may be submitted by a regular submit **<input>** element, by an image **<input>** element, or by a **<button>** element.

- Logo images can be added to the button face by enclosing an **<img>** element between **<button>** **</button>** tags.

- An option list is created by enclosing a number of **<option>** elements between **<select>** **</select>** tags.

- Option lists can be enclosed between **<optgroup>** **</optgroup>** tags that can specify an option group name to a **label** attribute.

- An **<input>** tag can include a **list** attribute to associate it with the **id** of a **<datalist>** element to provide pre-defined options.

- Each form control can be enclosed between **<label>** **</label>** tags to visually group them with text for styling purposes.

8 Paint on Canvas

Generate Canvas

In HTML, the **<canvas> </canvas>** tags create a bitmap canvas area on the page in which script can paint shapes and text. This can be used to dynamically generate graphs, game graphics, and visual images.

Initially, the canvas area is invisible and will, by default, be 300 pixels wide and 150 pixels high. Alternative dimensions can be specified to the **<canvas>** element's **width** and **height** attributes.

Optionally, fallback text can be included between the **<canvas>** **</canvas>** tags that will only be displayed by the browser in the event that the canvas area cannot be created.

In order to use the canvas, a script must first create a "CanvasRenderingContext2D" object. This snappily-named context object provides all the methods and properties needed to paint shapes and text in the canvas area. The context object is created using a **getContext( )** method of the canvas itself – for example, for a **<canvas>** element with an **id** of "canvas", like this:

```
const canvas = document.getElementById( 'canvas' )
const context = canvas.getContext( '2d' )
```

It is, however, good practice to first test for the existence of the canvas's **getContext( )** method before creating the context object:

```
const canvas = document.getElementById( 'canvas' )
if ( canvas.getContext )
{
  const context = canvas.getContext( '2d' )
}
```

Calls to the context object's methods, and assignments to its properties, can subsequently be added inside the **{ }** braces – to be implemented when the test succeeds.

A context object's **fillStyle** property can be assigned a color with which to paint a shape. For example, the context object's **fillRect( )** method can be called to paint a rectangle with the assigned color. This method requires four comma-separated "arguments" within its **( )** parentheses – to specify the XY coordinate position on the canvas of the top-left corner of the rectangle, its width, and its height:

context.**fillRect** (*x* , *y*, *width* , *height*)

Beware

JavaScript is case-sensitive so be sure to capitalize the **fillRect** method name correctly – with a capital "R".

Don't forget

Canvas coordinates have their XY origin at their top-left corner – so **fillRect(100,10, 50,50)** would paint a 50 pixel square 100 pixels from the left edge of the canvas and 10 pixels down from its top edge.

1 Create an HTML document that incorporates a canvas area of a specified size in the body section

```
<canvas id="canvas" width="500" height="300">

[Fallback Message]

</canvas>
```

canvas.html

2 Add a script with a function to paint the entire canvas area a specified color as soon as the document has loaded

```
<script>
( function ( ) {

  const canvas = document.getElementById( 'canvas' )
  if ( canvas.getContext )
  {
    const ctx = canvas.getContext( '2d' )
    ctx.fillStyle = 'Tomato'
    ctx.fillRect ( 0, 0, canvas.width, canvas.height )
  }

} ) ( )
</script>
```

Name the context object constant **ctx** for brevity as it will be typed often.

163

3 Save the HTML document, then open the web page in your browser to see the painted canvas area

Paint Shapes

The three basic shapes that can be painted on an HTML canvas are rectangle, circle, and polygon. A rectangle is the simplest shape to create using the context object's **fillRect( )** method, introduced in the previous example on pages 162-163, that has this syntax:

*ctx.***fillRect (*x, y, width, height*)**

Creating circles and polygons requires a little more effort, as they must both be initially created as a "path" describing the shape. The path of a circle describes the coordinates of its center point and its radius, and the path of a triangle describes the coordinates of each of its corners. A path always begins with a call to the context object's **beginPath( )** method – announcing the creation of a path.

A circular path is created with the context object's **arc( )** method, whose arguments first describe the coordinates of its center point and its radius. Additionally, because this method can also be used to create a partial circle, further arguments describe the start angle, end angle, and the direction in which to paint:

*ctx.***arc (*x, y, radius, startAngle, endAngle, direction*)**

Sadly, the start and end angles must be specified in "radians", rather than degrees, but degrees can easily be converted to radians using the expression *degrees****Math.PI/180**. A complete circle of 360 degrees can, therefore, start at zero and end at **360*Math.PI/180**, more simply expressed as just **Math.PI*2**. The final argument to the context object's **arc( )** method, describing the direction in which to paint, is a Boolean value of either **true** or **false** – where **false** is clockwise and **true** is counterclockwise.

A polygonal path is created with a context object's **moveTo( )** and **lineTo( )** methods that both require two XY coordinate arguments. Initially, the **moveTo( )** method describes the point at which to begin the path, like lifting a pen off paper and moving to a new point at which to begin drawing. Successive calls to a **lineTo( )** method then describe each corner point along the edge of the shape. Finally, a call to the context object's **closePath( )** method completes the path shape by returning to its starting point.

After creating a circular or polygonal path, a simple call to the context object's **fill( )** method will paint the shape with the color specified to the context object's current **fillStyle** property.

...cont'd

1 Create an HTML document containing a canvas
```
<canvas id="canvas" width="500" height="120">
[Fallback Message]</canvas>
```

shapes.html

2 Begin a script with a function to paint a 100-pixel square as soon as the document has loaded
```
( function ( ) {
  const canvas = document.getElementById( 'canvas' )
  if ( canvas.getContext )
  {
    const ctx = canvas.getContext( '2d' )
    ctx.fillStyle = 'Tomato'
    ctx.fillRect ( 75, 10, 100, 100 )
    /*  More instructions go here. */
  }
} ) ( )
```

3 Next, insert instructions to paint a 50-pixel radius circle
```
ctx.fillStyle = 'LawnGreen'
ctx.beginPath( )
ctx.arc( 275, 60, 50, 0, Math.PI*2, true )
ctx.fill( )
```

4 Now, insert instructions to paint a 100-pixel tall triangle
```
ctx.fillStyle = 'RoyalBlue'
ctx.beginPath( )
ctx.moveTo( 375, 110 )
ctx.lineTo( 425, 10 )
ctx.lineTo( 475, 110 )
ctx.closePath( )
ctx.fill( )
```

5 Save the HTML document, then open it in your browser to see the shapes get painted on the canvas

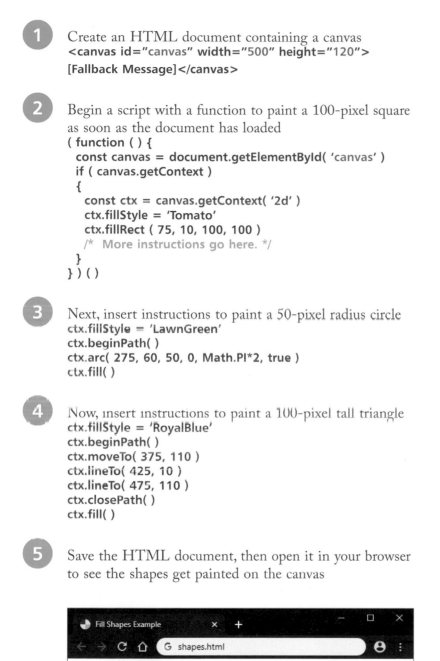

The blue border shown here has 50% fill transparency to illustrate the overlap – setting fill transparency is described in the next example on pages 168-169.

Stroke Borders

Just as a context object provides a **fillStyle** property, plus **fillRect( )** and **fill( )** methods that can be used to paint a shape, it also provides a **strokeStyle** property, plus **strokeRect( )** and **stroke( )** methods that can be used to paint borders.

A context object's **strokeStyle** property can be assigned a color with which to paint the borders. The context object's **strokeRect( )** method requires four comma-separated arguments within its ( ) parentheses – to specify the XY coordinate position on the canvas of the top-left corner of a rectangle, its width, and its height:

ctx.**strokeRect (*x, y, width, height*)**

Circular and polygonal borders are painted by first creating a path, as described in the previous example on pages 164-165, then calling the context object's **stroke( )** method.

Before painting a border, however, it is necessary to first specify a numeric pixel value to the context object's **lineWidth** property – to determine what brush width to use.

It should be noted that the border is painted of the specified **lineWidth** <u>centered</u> on the path – half outside and half inside. For example, with a **lineWidth** value of 20 pixels, the border gets painted with 10 pixels on each side of the path:

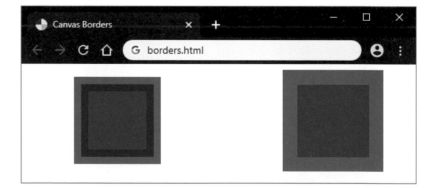

In order to paint a border entirely outside an existing fill path, the stroke path width and height need to be increased by the **lineWidth** value, and the XY coordinates need to be reduced by half the **lineWidth** value. For example, where the **lineWidth** value is 20 and the fill is created by **ctx.fillRect(20, 20, 100, 100)**, an external border is created by **ctx.strokeRect(10, 10, 120, 120)**.

1 Create an HTML document containing a canvas

```
<canvas id="canvas" width="500" height="120">
[Fallback Message]</canvas>
```

stroke.html

2 Begin a script with a function to paint a 6-pixel wide border centered on a 100-pixel square path

```
( function ( ) {
  const canvas = document.getElementById( 'canvas' )
  if ( canvas.getContext )
  {
    const ctx = canvas.getContext( '2d' )
    ctx.lineWidth = 6
    ctx.strokeStyle = 'Tomato'
    ctx.strokeRect ( 75, 10, 100, 100 )
    /*  More instructions go here. */
  }
} ) ( )
```

3 Next, insert instructions to paint a circular border

```
ctx.strokeStyle ='LawnGreen'
ctx.beginPath( )
ctx.arc( 275, 60, 50, 0, Math.PI*2, true )
ctx.stroke( )
```

4 Now, insert instructions to paint a triangular border

```
ctx.strokeStyle = 'RoyalBlue'
ctx.beginPath( )
ctx.moveTo( 375, 110 )
ctx.lineTo( 425, 10 )
ctx.lineTo( 475, 110 )
ctx.closePath( )
ctx.stroke( )
```

5 Save the HTML document then open it in your browser to see the borders get painted on the canvas

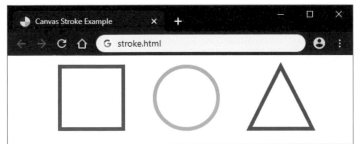

Fill Options

Fill and stroke colors can be specified to a context object's **fillStyle** and **strokeStyle** properties as a recognized color name, such as **Red**, or as a hexadecimal value, such as **#FF0000** or its shorthand equivalent **#F00**. Additionally, colors can be specified by stating their red, green, and blue component values in the range 0-255 with an **rgb( )** expression. For example, the color red can be expressed as **rgb(255, 0, 0)** – having maximum red component value but no green or blue. The fill color's transparency can also be specified in the range 0.0-1.0 as the fourth argument in an **rgba( )** expression. For example, the expression **rgba(255, 0,0, 0.5)** specifies maximum red of 50% transparency.

Hot tip

Should you want to paint a lot of shapes with the same fill transparency, you can specify a value in the range of 0.0-1.0 to the context object's **globalAlpha** property – for example, to set a 50% transparency with **ctx**.globalAlpha = 0.5.

Alternatively, a context object's **fillStyle** and **strokeStyle** properties can be assigned a "CanvasGradient" object that defines a multi-color gradient fill, in which one color gradually changes to another. To create a CanvasGradient object, the context object provides two methods. The **createLinearGradient( )** method requires four arguments to specify two XY coordinates at which to start and end a linear gradient. For example, **createLinearGradient(0,0, 100, 100)** defines a diagonal gradient from top left to bottom right. The **createRadialGradient( )** method requires six arguments to specify two XY coordinates and two radius values at which to start and end a radial gradient. For example, **createRadialGradient(50, 50, 10,50, 50, 100)** defines a radial gradient between two circles centered at the same point.

Each CanvasGradient object has an **addColorStop( )** method that requires two arguments to specify the position along the gradient in the range 0.0-1.0, and the color to paint at that position. For example, **addColorStop(0, 'Red')** begins the gradient fill with red, and **addColorStop(1, 'Blue')** ends the gradient fill with blue.

Don't forget

Do not simply specify the URL of an image as the first argument to the **createPattern( )** method.

A context object's **fillStyle** and **strokeStyle** properties can alternatively be assigned a "CanvasPattern" object that defines a pattern image and how it should be repeated. To create a CanvasPattern object, the context object provides a **createPattern( )** method. This requires two arguments specifying a loaded Image object and a repetition value of "repeat-x" (horizontal), "repeat-y" (vertical), or "repeat" (both). The Image object is created using the JavaScript **new Image( )** constructor and the URL of an image to be used by the pattern to its **src** property. The pattern can then be applied after the image has loaded.

...cont'd

1 Create an HTML document containing a canvas
```
<canvas id="canvas" width="500" height="120">
[Fallback Message]</canvas>
```

options.html

2 Begin a script with a function to paint colored rectangles
```
( function ( ) {
  const canvas = document.getElementById( 'canvas' )
  if ( canvas.getContext )
  {
    const ctx = canvas.getContext( '2d' )
    ctx.fillStyle = 'rgb( 255, 50, 0 )'
    ctx.fillRect ( 10, 10, 80, 80 )
    ctx.fillStyle = 'rgba( 0, 150, 255, 0.5 )'
    ctx.fillRect( 60, 30, 80, 80 )
    /* More instructions go here. */
  }
} ) ( )
```

3 Next, insert instructions to paint option-filled rectangles
```
const linear = ctx.createLinearGradient( 0, 10, 0, 110 )
linear.addColorStop( 0, 'Yellow' )
linear.addColorStop( 1, 'LawnGreen' )
ctx.fillStyle = linear ; ctx.fillRect( 150, 10, 100, 100 )

const radial = ctx.createRadialGradient(320, 60, 0, 320, 60, 70)
radial.addColorStop( 0, 'Yellow' )
radial.addColorStop( 1, 'LawnGreen' )
ctx.fillStyle = radial ; ctx.fillRect( 270, 10, 100, 100 )

const image = new Image( ) ; image.src = 'options.png'
image.onload = function( ) {
const pattern = ctx.createPattern( image, 'repeat' )
ctx.fillStyle = pattern ; ctx.fillRect( 390, 10, 100, 100 ) }
```

options.png 32px x 32px

4 Save the HTML document, then open it in your browser
to see the fill options

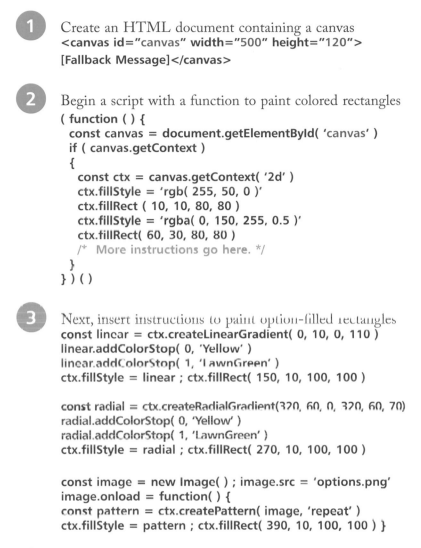

Paint Text

Text can be written on an HTML canvas using the context object's **fillText( )** and/or **strokeText( )** methods. These each require three arguments to specify the text to be written, enclosed within quote marks, and the XY coordinates at which to place the bottom-left corner of the text on the canvas. The text will appear painted in the canvas object's current **fillStyle** and/or **strokeStyle**.

The font in which to write the text can first be specified to the context object's **font** property. This accepts values in the same serial format as the CSS font shorthand property – to specify the font style, weight, size, and family. For example, the instruction **ctx.font = 'italic bold 90px Fantasy'** specifies an italic style, bold weight, 90-pixel size, and the "Fantasy" font family. Optionally, each value may be omitted from the series, so to simply specify an italic font style, the instruction could read **ctx.font='italic'**.

The context object can add shadow effects to any shape it paints, but is particularly useful to add drop-shadow effects to text. Shadow color, offset, and blur are specified by the context object's **shadowColor**, **shadowOffsetX**, **shadowOffsetY**, and **shadowBlur** properties. Positive **shadowOffsetX** values position the shadow to the right of the text, and positive **shadowOffsetY** values position the shadow below the text. Negative values may be specified to position the shadow to the left and above the text respectively:

The XY coordinates specified to **fillRect( )** position the top-left corner of a rectangle – but those specified to **fillText( )** position the text's bottom-left corner.

write.html

1. Create an HTML document containing a canvas
```
<canvas id="canvas" width="500" height="120">
[Fallback Message]</canvas>
```

2. Begin a script with a function to paint some bold text
```
( function ( ) {
  const canvas = document.getElementById( 'canvas' )
  if ( canvas.getContext )
  {
    const ctx = canvas.getContext( '2d' )
    ctx.font = 'bold 70px Arial, sans-serif'
    ctx.fillStyle = 'Tomato'
    ctx.fillText( 'HTML', 10, 60 )
    /*  More instructions go here. */
  }
} ) ( )
```

...cont'd

3   Next, insert instructions to paint some regular text

```
ctx.font = '25px Arial'
ctx.fillStyle = 'RoyalBlue'
ctx.fillText( 'with Context 2D', 10, 120 )
```

4   Now, insert instructions to paint outlined text

```
ctx.font = 'italic bold 40px Fantasy'
ctx.strokeStyle = 'LawnGreen'
ctx.strokeText( 'Canvas Fun', 10, 95 )
```

5   Specify values to each context object shadow property

```
ctx.shadowOffsetX = 2
ctx.shadowOffsetY = 2
ctx.shadowBlur = 3
ctx.shadowColor ='Black'
```

6   Finally, insert instructions to paint text that will receive a drop-shadow effect

```
ctx.font = 'italic bold 60px Fantasy'
ctx.fillStyle = 'Yellow'
ctx.fillText( 'Drop', 250, 50 )
ctx.fillText( 'Shadow', 250, 100 )
```

7   Save the HTML document, then open it in your browser to see the text written on the canvas and admire the drop-shadow effect

Shadow properties are global, so once values have been specified to them, all further shapes painted onto the canvas will automatically receive a shadow. This can be prevented, however, by assigning a value of "transparent" to the **shadowColor** property.

Paint Lines

Lines are initially created as paths that start with the context object's **beginPath( )** method, then use its **moveTo( )** and **lineTo( )** methods to specify the coordinates on the HTML canvas at which to draw the line. A call to the context object's **stroke( )** method draws the line using the current **strokeStyle** and **lineWidth**. By default, the ends of the line are drawn flat, exactly abutting the canvas coordinates, but alternate line endings can be specified to the context object's **lineCap** property – a **round** value adds a semi-circular cap, and a **square** value adds a rectangular cap. Each cap adds half the line's width. Normal line endings can be resumed by specifying the default **butt** value.

Where two lines join at an angle, they automatically receive an extension beyond the specified coordinates to create a mitered point. This extends the outer edge of each line until they meet, then fills the triangle formed by the extension. When two lines join at a very acute angle, the extension needed to form the miter triangle can be lengthy, so the context object provides a **miterLimit** property to constrain the extension length. Initially, this property has a value of "10", which is generally desirable.

The extended miter normally created where two lines join can be prevented by specifying a "bevel" value to the context object's **lineJoin** property, or an attractive filled arc can be added to the line ends by specifying a "round" value. Normal line joints can be resumed by specifying the default "miter" value to the context object's **lineJoin** property.

lines.html

1. Create an HTML document containing a canvas
    ```
    <canvas id="canvas" width="500" height="120">
    [Fallback Message]</canvas>
    ```

2. Begin a script with a function to paint a triangle
    ```
    ( function ( ) {
      const canvas = document.getElementById( 'canvas' )
      if ( canvas.getContext )
      {
        const ctx = canvas.getContext( '2d' ) ; ctx.lineWidth = 20
        ctx.strokeStyle = 'Tomato' ; ctx.beginPath( )
        ctx.moveTo( 20, 110 ) ; ctx.lineTo( 70, 30 )
        ctx.lineTo( 110, 110 ) ; ctx.closePath( ) ; ctx.stroke( )
        /*  More instructions go here. */
      }
    } ) ( )
    ```

...cont'd

3 Next, insert instructions to paint lines with different ends
```
ctx.strokeStyle = 'LawnGreen'

ctx.beginPath( )
ctx.lineCap = 'butt'
ctx.moveTo( 160, 30 ) ; ctx.lineTo( 160, 110 ) ; ctx.stroke( )

ctx.beginPath( )
ctx.lineCap = 'round'
ctx.moveTo( 200, 30 ) ; ctx.lineTo( 200, 110 ) ; ctx.stroke( )

ctx.beginPath( )
ctx.lineCap = 'square'
ctx.moveTo( 240, 30 ) ; ctx.lineTo( 240, 110 ) ; ctx.stroke( )
```

4 Now, insert instructions to paint lines with different joints
```
ctx.strokeStyle = 'RoyalBlue'

ctx.beginPath( )
ctx.lineJoin = 'miter'
ctx.moveTo( 300, 80 ) ; ctx.lineTo( 330, 30 )
ctx.lineTo( 330, 120 ) ; ctx.stroke( )

ctx.beginPath( )
ctx.lineJoin = 'round'
ctx.moveTo( 370, 80 )  ; ctx.lineTo( 400, 30 )
ctx.lineTo( 400, 120 )  ; ctx.stroke( )

ctx.beginPath( )
ctx.lineJoin = 'bevel'
ctx.moveTo( 440, 80 ) ; ctx.lineTo( 470, 30 )
ctx.lineTo( 470, 120 ) ; ctx.stroke( )
```

5 Save the HTML document, then open it in your browser
to see the lines painted on the canvas, and admire their
ends and joints

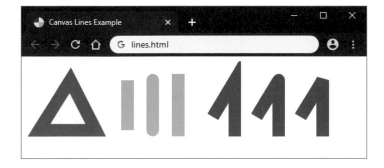

Don't forget

Lines are painted by
stroking, but when
closePath() has been
used to create a shape,
that shape can be filled.

Paint Curves

The context object's **arc( )** method, which was used in the example on pages 164-165 to paint a complete circle, can be used to paint a curved line on the canvas that is simply part of a circle's circumference. Recall that the **arc()** method requires six arguments to specify the circle's center point, radius, start angle (in radians), end angle (in radians), and the direction in which to paint – with this syntax:

ctx.arc (x, y, radius, startAngle, endAngle, direction)

When painting a complete circle, the Boolean direction value is irrelevant, but it is important when painting an arc to determine which part of the circumference is to be painted. For example, when the start angle is at 3 o'clock (zero) and the end angle is at 12 o'clock (**Math.PI/180 * 270**), painting counterclockwise (**true**) creates an arc that is only one quarter of the circumference, but clockwise (**false**) creates an arc of the other three quarters.

Simple curved lines that are not arcs can be painted on the canvas using the context object's **quadraticCurveTo( )** method. This requires four arguments – specifying the XY coordinates of <u>one</u> invisible control point, and the XY coordinates of the line end point.

Complex curved lines can be painted on the canvas using the context object's **bezierCurveTo( )** method. This requires six arguments – specifying the XY coordinates of <u>two</u> invisible control points, and the XY coordinates of the line end point. In each case, the line swerves towards the control points to form the curve:

Hot tip

The formula to convert degrees to radians is described on page 164.

HTML

curves.html

1. Create an HTML document containing a canvas
```
<canvas id="canvas" width="500" height="120">
[Fallback Message]</canvas>
```

2. Begin a script with a function to paint two 100-pixel square rectangles – upon which to paint arcs
```
( function ( ) {
  const canvas = document.getElementById( 'canvas' )
  if ( canvas.getContext )
  {
    const ctx = canvas.getContext( '2d' )
    ctx.fillStyle = 'Yellow'
    ctx.fillRect( 70, 10, 100, 100 )
    ctx.fillRect( 200, 10, 100, 100 )
    /*  More instructions go here. */
  }
} ) ( )
```

3. Next, insert instructions to paint two arcs from the same circumference position, but painted in different directions

```
ctx.lineWidth = 15
ctx.strokeStyle = 'Tomato'

ctx.beginPath( )
ctx.arc( 70, 60, 50, 0, Math.PI/180*90, true )
ctx.stroke( )

ctx.beginPath( )
ctx.arc( 120, 60, 50, 0, Math.PI/180*90, false )
ctx.stroke( )
```

The background rectangles in this example are included to illustrate positioning – paint 1-pixel square rectangles at the control point coordinates (350,100), (550,10), and (450,100) to see how they swerve the curves.

4. Now, insert instructions to paint a filled semi-circle

```
ctx.beginPath( )
ctx.arc( 250, 60, 50, Math.PI/180*90, Math.PI/180*270, true )
ctx.fillStyle = 'LawnGreen' ; ctx.fill( )
```

5. Paint a simple curve using one control point

```
ctx.strokeStyle = 'RoyalBlue'
ctx.beginPath( )
ctx.moveTo( 340, 10 )
ctx.quadraticCurveTo( 340, 100, 420, 100 )
ctx.stroke( )
```

6. Now, paint a complex curve using two control points

```
ctx.beginPath( )
ctx.moveTo( 400, 10 )
ctx.bezierCurveTo( 500, 10, 400, 100, 500, 100 )
ctx.stroke( )
```

7. Save the HTML document, then open the web page in your browser to see the rectangles, arcs, and curves painted on the canvas

Translate Coordinates

The canvas XY coordinate origin, which by default sets 0,0 at the top-left corner of the canvas, can be changed by specifying new XY coordinate origins to the context object's **translate( )** method. After painting at the new coordinates, the canvas state can then be saved using the context object's **save( )** method, before restoring the initial default origin using the context object's **restore( )** method. This technique is especially useful when painting multiple shapes from a script loop to move the canvas on each iteration.

Script loops can also scale shapes on successive iterations using the context object's **scale()** method. This requires two arguments to specify the "scale factor" in the horizontal and vertical directions. For example, **ctx.scale(0.5, 0.5)** scales down by 50% in each direction, and **ctx.scale(1.5, 1.5)** scales up by 50%.

The context object also provides a **rotate( )** method that allows the canvas to be rotated clockwise by the angle (expressed in radians) specified as its single argument. For example, specifying an argument of **Math.Pi*2/36** rotates the canvas 10 degrees (360/36). Script loops can call the context object's **rotate( )** method on successive iterations of a loop to paint shapes in a circular pattern:

The formula to convert degrees to radians is described on page 164.

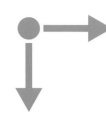

translate.html

1 Create an HTML document containing a canvas
```
<canvas id="canvas" width="500" height="120">
[Fallback Message]</canvas>
```

2 Begin a script with a function to paint square rectangles by translating the XY origin on each iteration of a loop
```
( function ( ) {
  const canvas = document.getElementById( 'canvas' )
  if ( canvas.getContext )
  {
    const ctx = canvas.getContext( '2d' )
    let i, j = 0 ; ctx.fillStyle = 'Tomato'
    for( i = 0 ; i < 3 ; i++ )
    {
      for( j = 0 ; j < 3 ; j++ )
      {
          ctx.save( ) ; ctx.translate( 40*j, 40*i )
          ctx.fillRect( 0, 0, 30, 30 ) ; ctx.restore( )
      }
    }
    /*  More instructions go here. */
  }
} ) ( )
```

3 Next, insert a loop construct to paint a series of rectangles – by translating the XY canvas origin and scaling down on each iteration of the loop

```
ctx.fillStyle = 'LawnGreen'
ctx.translate( 130, 0 )
for( i = 0 ; i < 3 ; i++ )
{
  ctx.fillRect( 0, 0, 100, 100 )
  ctx.translate( 110, 0 )
  ctx.scale( 0.75, 0.75 )
}
```

Beware

The context accumulates calls to **translate( )**, **scale( )**, and **rotate( )** in a matrix – so after rotating, say, 45 degrees, a subsequent call to **translate( )** along the X axis will move diagonally, not horizontally! The next example on pages 178-179 shows how to avoid this by resetting the matrix.

4 Now, insert instructions to paint a series of circles – by rotating the canvas on each iteration of a loop

```
ctx.fillStyle = 'RoyalBlue'
ctx.translate( 150, 120 )
for ( i = 1 ; i < 6 ; i++ )
{
  for ( j = 0 ; j < i*6 ; j++ )
  {
    ctx.rotate( Math.PI*2 / ( i*6 ) )
    ctx.beginPath( )
    ctx.arc( 0, i*22.5, 8, 0, Math.PI*2, true )
    ctx.fill( )
  }
}
```

177

5 Save the HTML document then open the web page in your browser to see the shapes painted at the various coordinates by the loops

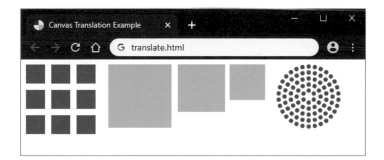

Transform Shapes

When creating shapes and paths, the context object applies a "transformation matrix" to the canvas coordinates. Initially, this provides an "identity transform" for the context object's methods. For example, it ensures rectangle corner angles are all 90 degrees.

The transformation matrix can be manipulated, however, to allow the context object's methods to behave differently – for example, to create skewed rectangles that do not have 90-degree corners.

Calling a context object's **transform( )** method can apply a transformation by multiplying the current matrix values. For example, the call **ctx.transform(1, -0.3, 0, 1, 0, 0)** will skew rectangles when painting. The current matrix retains previous transformations but, usefully, can be reset to the default identity matrix with the call **ctx.setTransform(1, 0, 0, 1, 0, 0)** – so previous calls to **rotate( )**, **scale( )**, **translate( )**, and **transform( )** are ignored.

The appearance of shapes can be also be modified by first defining a clipping path to act as a mask. Subsequently, only shapes, or parts of shapes, that are inside the clipping path will be painted. A clipping path is simply a specified path – created like any other path – that ends with a call to the context object's **clip( )** method:

transform.html

1 Create an HTML document containing a canvas
```
<canvas id="canvas" width="500" height="150">
[Fallback Message]</canvas>
```

2 Begin a script with a function to paint a 100-pixel square rectangle – using the identity transformation matrix
```
( function ( ) {
  const canvas = document.getElementById( 'canvas' )
  if ( canvas.getContext )
  {
    const ctx = canvas.getContext( '2d' )
    ctx.fillStyle = 'Tomato'
    ctx.fillRect( 20, 40, 100, 100 )
    /*  More instructions go here. */
  }
} ) ( )
```

3 Next, insert instructions to paint a 100-pixel square rectangle – by multiplying the current matrix values
```
ctx.fillStyle = 'LawnGreen'
ctx.transform( 1, -0.3, 0, 1, 0, 0 )
ctx.fillRect( 160, 90, 100, 100 )
```

4 Now, insert an instruction to reset to the default identity
transformation matrix – forgetting the last transformation
ctx.setTransform(1, 0, 0, 1, 0, 0)

5 Paint another rectangle – once more using the identity
transformation matrix
ctx.fillStyle = 'RoyalBlue'
ctx.fillRect(350, 10, 130, 130)

6 Next, create a circular clipping path, centered in the
rectangle just painted
ctx.beginPath( )
ctx.arc(415, 75, 50, 0, Math.PI*2, true)
ctx.clip( )

7 Now, paint two more rectangles over the clipping path
ctx.fillStyle – 'Yellow'
ctx.fillRect(350, 10, 130, 130)

ctx.fillStyle = 'Purple'
ctx.fillRect(350, 10, 65, 65)

8 Save the HTML document, then open it in your browser
to see the skewed rectangle painted using the modified
transformation matrix, and rectangles clipped by the mask

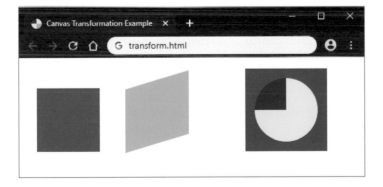

You can edit the
transform( ) argument
list to skew along the
other axis instead with
(1, 0, -0.3, 1, 0, 0).

Consider each canvas
element to have, by
default, a clipping path
that is the same size as
the entire canvas – so no
clipping occurs.

Animate Canvas

Animations can easily be created on a canvas simply by repeatedly clearing the current canvas then repainting it with shapes at a modified position. A canvas can be repainted faster than the human eye can detect, so animations appear to be very smooth. The fundamental components of a canvas animation script are:

● Initialize a context object and shape starting positions.

● Clear the canvas, then paint shapes onto the canvas.

● Calculate new shape positions for the next repaint.

Where the canvas animation has a static background and border, these can be created as styles so they need not be repeatedly painted onto the canvas:

animate.html

 Create an HTML document containing a canvas
```
<canvas id="canvas" width="500" height="150">
[Fallback Message]</canvas>
```

② Add a style sheet to color the canvas
```
<style>
canvas { border:2px solid Purple; background: Moccasin; }
</style>
```

③ Begin a script that initializes a constant and four variables to specify an initial XY coordinate position, and to specify a horizontal and vertical step size of five pixels
```
const canvas = document.getElementById( 'canvas' )
let x = 20, y = 130
let dx = dy = 5
```

④ Now, add a function that affirms the canvas size and initializes a context object property
```
function paint( ) {

  if ( canvas.getContext )
  {
    canvas.width = 500 ; canvas.height = 150
    const ctx = canvas.getContext( '2d' )

    /*  More instructions go here. */
  }
}
```

5 Next, insert instructions to paint a circle at the current stored XY coordinates position

```
ctx.beginPath( )
ctx.arc( x, y, 20, 0, Math.PI*2, true )
ctx.closePath( )
ctx.fillStyle = 'Purple'
ctx.fill( )
```

6 Then, insert instructions to calculate new XY coordinates and store them in the global variables

```
if ( x < 20 || x > 480 ) { dx = -dx }
if ( y < 20 || y > 130 ) { dy = -dy }
x += dx
y += dy
requestAnimationFrame( paint )
```

7 Finally, after the end of the function block, add an instruction to start the animation when the page loads

```
requestAnimationFrame( paint )
```

8 Save the HTML document, then open it in your browser to see an animated "ball" bounce around the canvas

Hot tip

Notice how the polarity of the direction step gets reversed when the ball collides with a perimeter – so it doesn't bounce right off the canvas.

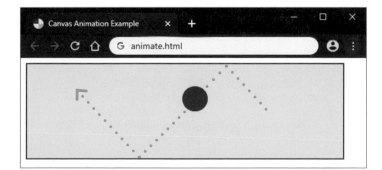

Hot tip

The call to the **requestAnimationFrame( )** method will usually repaint the canvas 60 times per second to match the display refresh rate of most web browsers.

Draw Pictures

Images can be drawn on a canvas alongside painted shapes. The script must first create a new **Image** object, then the path to the image file must be assigned to that object's **src** property. Finally, the image can be drawn onto the canvas by a call to the context object's **drawImage( )** method to specify the image source and XY coordinates at which to draw the image on the canvas.

A canvas can become interactive by adding event listeners to the entire canvas, or to individual items on the canvas by specifying XY coordinate boundaries for the event listener. The script must, however, recognize the distance the canvas is offset from the top-left corner of the window to correctly locate the interactive item.

banner.html

① Create an HTML document containing a canvas and a style sheet to color the canvas
```
<canvas id="canvas" width="500" height="150">
[Fallback Message]</canvas>

<style>
canvas { border:2px solid Purple; background: Moccasin; }
</style>
```

② Begin a script that initializes a constant and nominates an event listener function for the click event
```
const canvas = document.getElementById( 'canvas' )
canvas.addEventListener( 'click', position )
```

③ Next, create an **Image** object for an image that will be drawn on a canvas
```
const python = new Image( )
python.src = 'banner-python.png'
```

banner-python.png
130px x 130px

④ Now, add a function that affirms the canvas size and initializes a context object property
```
function paint( ) {

  if ( canvas.getContext )
  {
    canvas.width = 500 ; canvas.height = 150
    const ctx = canvas.getContext( '2d' )

    /*  More instructions go here. */
  }
}
```

5 Insert instructions to draw the image and paint a rectangle displaying some text

```
ctx.drawImage( python, 10, 10 )

ctx.fillStyle = 'YellowGreen'
ctx.fillRect( 350, 20, 130, 110 )

ctx.fillStyle = 'White'
ctx.font = 'bold 20px Arial, sans-serif'
ctx.fillText( 'Click For', 370, 70 )
ctx.fillText( 'Details', 380, 100 )
```

6 After the end of the function block, add the event listener function for the click event

```
function position( event ) {

  const x = ( event.x - canvas.offsetLeft )
  const y = ( event.y - canvas.offsetTop )
  if( x > 350 && x < 480 )
  {
    if( y > 20 && y < 130 )
    {
      location='http://www.ineasysteps.com'
    }
  }
}
```

Hot tip

The click event returns XY coordinates of the cursor position in the window, so you can subtract the offset distance of the canvas to find the XY coordinates of the cursor position within the canvas.

7 Finally, after the end of both function blocks, add an instruction to draw a banner when the page loads

```
onload = paint
```

8 Save the HTML document, then open it in your browser to see the image and interactive canvas area

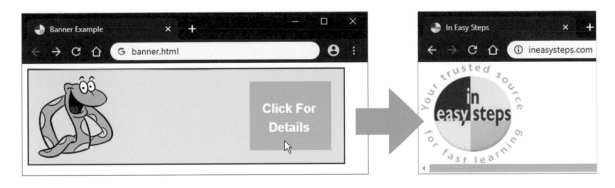

Animate Banners

An HTML canvas can combine animation with painted shapes and text, drawn images, and components that recognize events, to create compelling visual effects, interactive games, data visualization, photo manipulation, and much more. For example, a canvas is commonly used for the animated advertising banners you often see on web pages.

 Create a copy of the previous example **banner.html** (on pages 182-183) and rename it as "banner-anim.html"

 Begin to edit the script by adding a statement to initialize two variables at the very start of the script
let count = 0, y = 180

 After the existing statements within the **paint( )** function, insert statements to add static text in the middle center of the canvas
ctx.fillStyle = 'Purple'
ctx.font = 'bold 50px Arial, sans-serif'
ctx.fillText('Python', 150, 90)

 Next, insert statements to add text that will pop up in the top center of the canvas during an animation sequence
if(count > 100) {
  ctx.fillStyle = 'Black'
  ctx.font = 'bold 25px Arial, sans-serif'
  ctx.fillText('Learn to Code', 150, 40)
}

 Now, insert statements to add text that will scroll up in the bottom center of the canvas during the animation
if(count > 200) {
  ctx.fillStyle = 'Purple'
  ctx.font = 'bold 27px Arial, sans-serif'
  ctx.fillText('in easy steps', 154, y)
  if (y > 120) { y = (y - 0.5) }
}

 Then insert statements to increment the counter during the animation and repeat the sequence
count++
if(count > 500) { count = 0 ; y = 180 }

Finally, insert a statement to perform the animation
requestAnimationFrame(paint)

8 Save the HTML document, then open it in your browser
to see the animated interactive banner

Count is 0-100.

Count is 101-200.

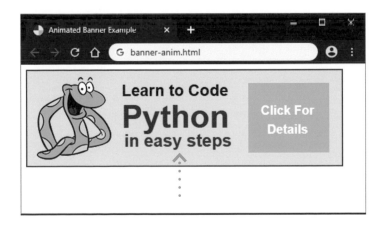

Count is 201-500.

Summary

- The HTML **<canvas>** element creates a bitmap canvas area on a page in which JavaScript can paint shapes and text – using the canvas context object's methods and properties.

- A context object's **fillStyle**, **strokeStyle**, and **lineWidth** properties specify the current fill color, stroke color, and line width.

- A rectangle can be simply painted using the context object's **fillRect( )** or **strokeRect( )** methods to specify position and size.

- Creating a path always begins by calling the context object's **beginPath( )** method, and may be closed with **closePath( )**.

- A path may use the context object's **moveTo( )**, **lineTo( )**, and **arc( )** methods to describe path coordinates, then paint the path on the canvas by calling its **fill( )** or **stroke( )** methods.

- Fills may also be semi-transparent colors, gradients, or patterns.

- Text painted onto a canvas by the context object's **fillText( )** and **strokeText( )** methods can be enhanced by a shadow effect.

- A path can specify the appearance of line endings and line joints to the context object's **lineCap** and **lineJoin** properties.

- A context object's **quadraticCurveTo( )** and **bezierCurveTo( )** methods both specify end point XY coordinates and control point XY coordinates to define curves.

- The context object's **translate( )** method changes the canvas XY origin, its **rotate( )** method rotates the canvas, and its **scale( )** method modifies a shape by a specified scale factor.

- The canvas state can be saved by the context object's **save( )** method, and the default origin restored by its **restore( )** method.

- The context object's **transform( )**, **setTransform( )**, and **clip( )**. methods can be used to modify its painting behavior.

- The **drawImage( )** method is used to add images onto a canvas.

- Canvas animation is achieved by repeatedly repainting the canvas faster than the human eye can detect, by calling the **requestAnimationFrame( )** method.

Index

189

Q

R

S